AF870883

Response Surface Methodology - Research Advances and Applications

Edited by Palanikumar Kayarogannam

Published in London, United Kingdom

Response Surface Methodology - Research Advances and Applications
http://dx.doi.org/10.5772/intechopen.102317
Edited by Palanikumar Kayarogannam

Contributors
Sheriff Lamidi, Nurudeen Olaleye, Yakub Bankole, Aishat Obalola, Emmanuella Aribike, Idris Adigun, Nam-Ky Nguyen, Mai Phuong Vuong, John J. Borkowski, Sawsan Mahmood, Ali Ali, Wissam Zam, Ayhem Darwesh, Julio Romero-Noguera, Nuria Pérez-Villares, Fernando Bolívar-Galiano, Rafael Bailón-Moreno, Samuel K. Kalu Otisi, Chigoziri N. Nnaemeka Njoku, Aysun Sagbas, Asif Ahmad, Shahnawaz Alam, Meenu Sharma, Palanikumar Kayaroganam

© The Editor(s) and the Author(s) 2023
The rights of the editor(s) and the author(s) have been asserted in accordance with the Copyright, Designs and Patents Act 1988. All rights to the book as a whole are reserved by INTECHOPEN LIMITED. The book as a whole (compilation) cannot be reproduced, distributed or used for commercial or non-commercial purposes without INTECHOPEN LIMITED's written permission. Enquiries concerning the use of the book should be directed to INTECHOPEN LIMITED rights and permissions department (permissions@intechopen.com).
Violations are liable to prosecution under the governing Copyright Law.

Individual chapters of this publication are distributed under the terms of the Creative Commons Attribution 3.0 Unported License which permits commercial use, distribution and reproduction of the individual chapters, provided the original author(s) and source publication are appropriately acknowledged. If so indicated, certain images may not be included under the Creative Commons license. In such cases users will need to obtain permission from the license holder to reproduce the material. More details and guidelines concerning content reuse and adaptation can be found at http://www.intechopen.com/copyright-policy.html.

Notice
Statements and opinions expressed in the chapters are these of the individual contributors and not necessarily those of the editors or publisher. No responsibility is accepted for the accuracy of information contained in the published chapters. The publisher assumes no responsibility for any damage or injury to persons or property arising out of the use of any materials, instructions, methods or ideas contained in the book.

First published in London, United Kingdom, 2023 by IntechOpen
IntechOpen is the global imprint of INTECHOPEN LIMITED, registered in England and Wales, registration number: 11086078, 5 Princes Gate Court, London, SW7 2QJ, United Kingdom

British Library Cataloguing-in-Publication Data
A catalogue record for this book is available from the British Library

Additional hard and PDF copies can be obtained from orders@intechopen.com

Response Surface Methodology - Research Advances and Applications
Edited by Palanikumar Kayarogannam
p. cm.
Print ISBN 978-1-83880-288-2
Online ISBN 978-1-83880-299-8
eBook (PDF) ISBN 978-1-83880-463-3

We are IntechOpen, the world's leading publisher of Open Access books

Built by scientists, for scientists

6,300+
Open access books available

171,000+
International authors and editors

190M+
Downloads

156
Countries delivered to

Our authors are among the
Top 1%
most cited scientists

12.2%
Contributors from top 500 universities

WEB OF SCIENCE™

Selection of our books indexed in the Book Citation Index
in Web of Science™ Core Collection (BKCI)

Interested in publishing with us?
Contact book.department@intechopen.com

Numbers displayed above are based on latest data collected.
For more information visit www.intechopen.com

Meet the editor

Prof. Dr. Palanikumar is a professor and principal at Sri Sai Ram Institute of Technology, India. He is involved in teaching, research, development, and innovations in the field of higher technical education. He received the National-Visvesvaraya Best Teacher Award from the All India Council for Technical Education (AICTE), Government of India, in 2021. He is listed among the world's top 2% of materials researchers by Stanford University, USA. He has published more than 450 papers and 30 patents in engineering and technology. Dr. Palanikumar's areas of interest include statistical techniques, modeling and optimization, total quality management, composite materials, machining, and additive manufacturing.

Contents

Preface

Response surface methodology (RSM) is used to predict and optimize process parameters in science, engineering, and technology problems. Its applications are not limited to these areas, however; RSM has been used in many fields, including medicine, pharmaceuticals, and even the arts. This book explores the use of RSM for modeling and optimization in a variety of disciplines.

The book includes eight chapters with case studies to illustrate relevant concepts. Chapter 1 provides an introduction to the topic and basic experimental design methods. Chapter 2 discusses the application of central composite design and its effects on RSM. Chapter 3 highlights the application of RSM in product development and design. Chapter 4 discusses cyclic generators for Box–Behnken designs and develops new second-order designs. Chapters 5–8 examine modeling and optimization using RSM. Chapter 5 analyzes and optimizes bead geometry parameters in the welding process. Chapter 6 presents a case study on electrical discharge machining and uses RSM to optimize the process parameters. Chapter 7 discusses the selection of optimal conditions for baker's yeast for the production of grape juice. Finally, Chapter 8 applies RSM to fine arts and the restoration of paintings.

This volume is a useful resource for those interested in RSM and who wish to use it in new and innovative research.

Palanikumar Kayaroganam
Department of Mechanical Engineering,
Sri Sai Ram Institute of Technology,
Chennai, India

Section 1

Introduction to Response Surface Methodology

Chapter 1

Introductory Chapter: Response Surface Methodology

Palanikumar Kayaroganna m

1. Introduction

The development of empirical models and optimization is the focus of the mathematical and statistical methodologies, which is called as the response surface methodology (RSM). In RSM, the experiment optimizes a response (output variable) that is sensitive to many factors. As a result of it, study designs will accomplish these aims (input variables). Many trials are conducted using different variables, and the best variables are selected to accomplish the goals successfully.

RSM is the first model to incorporate observed responses [1]. Afterward, the numerical experimentation-based modeling technique are emerged. Also, several possible errors exist in the solution. Insufficient convergence of iterative methods, rounding mistakes, and a discrete representation of a continuous physical event are all potential sources of numerical noise in computer experiments. If the measurements are inaccurate in a scientific experiment, the conclusions will be affected at the end. If the measurements are inaccurate, then the findings of a physical experiment will be inaccurate [2–5], at last. As a result of it, the RSM methodology adopts the position where errors occur at random. When used for design optimization, RSM reduces the need for high-priced analytical approaches like the computational fluid dynamics (CFD) analysis.

For the purpose of optimizing designs, many researchers have discussed the use of RSM. For example, the optimum surface roughness on machining is optimized by determining the optimum cutting speed (x_1) and feed (x_2). It is observed that cutting speed and feed affect the surface roughness:

$$y = f(x_1, x_2) + \varepsilon \quad (1)$$

where ε denotes experimental error due to back ground noise, manual error, etc.

RSM is graphically displayed in either a two- or three-dimensional plot known as contour plot, and it is interpreted in two different ways. When all the factors are held at constant except for the x_i and x_j coordinates, the resulting curves are called contours. It specifies the highest reaction on each form's surface. **Figure 1** shows the steps in RSM, which clearly shows how the optimization is carried out by RSM.

IntechOpen

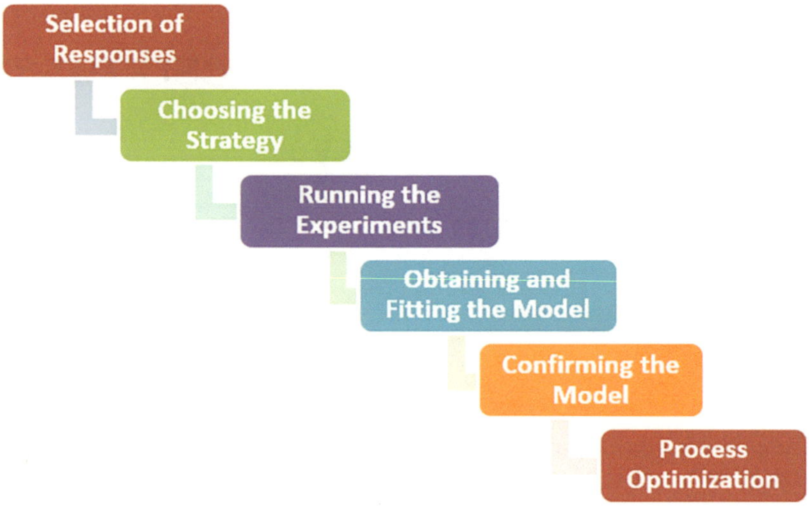

Figure 1.
Response surface methodology.

2. Approximate model function

The relationship between the response and a nonresponse variable is unclear. First, RSM determines the best near model. Further the vast majority of models used in practice are polynomials of low order (first or second order). When a structure of the model has curvature, it includes the interaction effects and square effects. This model is observed as:

$$y = \beta_0 + \sum_{i=1}^{k} \beta_i x_i + \sum_{i=1}^{k} \beta_{ii} x_i^2 + \sum_{i<j} \sum \beta_{ij} x_i x_j + \varepsilon \tag{2}$$

Variety of work has been carried out in response surface modeling. This degree of usefulness is quantified by the goodness-of-fit metric. Sensitivity data is used to reduce the number of computer simulation analyses for model fitting, though it is not always readily available or affordable.

Some of the important methods used for RSM are as follows:

- **Design of experiments:** RSM relies mainly on design of experiments, abbreviated as DoE [1]. This method is mainly developed to fit the models for physical experiment data. Also, it is used for numerical experiment data.

- **Full factorial design:** It is necessary to do a full factorial analysis [6] for creating approximate model in establishing the relationship between the design variables.

- **Central composite design (CCDs):** CCDs are considered, as the first-order designs with an extra center and axial points in order to estimate the tuning parameters of second-order models,

- **D-optimal designs:** The experimental points provide the most reliable estimations of the response model coefficients using quadratic models considering the D-optimality criterion [7].

- **Taguchi's contribution to experimental design:** Taguchi has used the orthogonal arrays for designing the experiment. Taguchi's design allows for fewer testing than a full factorial design.

Other approaches like [8–10] are also considered by some researchers. It is clearly known that RSM is an important method in statistical design, and it is applied for different kind of experiments. Research advances are carried out in different fields of Engineering, Science, and Technology [11]. The application of RSM in the different fields is innumerable. Hence, RSM method finds its applications in textile industry, food industries, chemical industries, mechanical industries, pharmaceutical industries, etc. By using proper design of experiments, the modeling and optimization are carried out without cumbersome efforts.

3. Conclusion

RSM is very much useful for analyzing and optimizing the experimental and numerical responses. It has got close approximations of both experimental and numerical results by choosing the correct method of experimentation. The best plans for experiments have the experimental locations dispersed around the area of interest. Recently, many new plans are implemented for analyzing the data. In spite of computer-based many modeling and optimization techniques are developed, RSM is an indispensable method and is used for modeling and optimization in the field of Science, Engineering, and Technology.

Author details

Palanikumar Kayarogannam
Department of Mechanical Engineering, Sri Sai Ram Institute of Technology, Chennai, India

*Address all correspondence to: palanikumar@sairamit.edu.in

IntechOpen

© 2023 The Author(s). Licensee IntechOpen. This chapter is distributed under the terms of the Creative Commons Attribution License (http://creativecommons.org/licenses/by/3.0), which permits unrestricted use, distribution, and reproduction in any medium, provided the original work is properly cited.

References

[1] George EP et al. Draper, Empirical Model-Building and Response Surfaces. New York: Wiley; 1987. p. 688

[2] Giunta AA et al. Wing Design for a High-Speed Civil Transport Using a Design of Experiments Methodology. Blacksburg, United States: Department of Computer Science, Virginia Polytechnic Institute & State University; 1996

[3] Audze P, Eglais V. New approach for planning out of experiments. Problems of Dynamics and Strengths. 1977;**35**:104-107

[4] Toropov V, Fred V, Markine V. Refinements in the multi-point approximation method to reduce the effects of noisy structural responses. In: 6th Symposium on Multidisciplinary Analysis and Optimization. Bellevue, WA, U.S.A. 1996

[5] Van Kampen A, Huiskes R. The three-dimensional tracking pattern of the human patella. Journal of Orthopaedic Research. 1990;**8**(3):372-382

[6] Montgomery DC. Response surface methods and other approaches to process optimization. In: Montgomery DC, editor. Design and Analysis of Experiments. New York: John Wiley & Sons; 1997. pp. 427-510

[7] Myers RH, Montgomery DC. Response Surface Methodology: Process and Product Optimization Using Designed Experiment. Hoboken, New Jersey, U.S.: John Wiley & Sons. Inc; 1995

[8] McKay MD et al. A comparison of three methods for selecting values of input variables in the analysis of output from a computer code. Technometrics. 1979;**21**(2):239-245

[9] Van Keulen F, Toropov VV. The multi-point approximation method in a parallel computing environment. Zeitschrift Fur Angewandte Mathematik Und Mechanik. 1999;**79**:S67-S70

[10] Venter JC, Smith HO, Hood L. A new strategy for genome sequencing. Nature. 1996;**381**(6581):364-366

[11] Palanikumar K. Introductory chapter: Response surface methodology in engineering science. In: Response Surface Methodology in Engineering Science. London, UK: IntechOpen; 2021. DOI: 10.5772/intechopen.100484

Section 2

Applications of Response Surface Methodology

Chapter 2

Application of Central Composite Design with Design Expert v13 in Process Optimization

Chigoziri N. Njoku and Samuel K. Otisi

Abstract

This chapter is focused on the study application of central composite design, in response surface methodology. We have reviewed this concept and applied it to optimize Biodiesel yield from transesterification of methanol and vegetable oil with a catalyst derived from eggshell using design expert 13. This optimization was carried out with reaction conditions of reaction time, methanol to oil ratio, catalyst loading, and reaction temperature. Data used as an instance was collected and analyzed from the work of Tshizanga et al. and the result obtained for a randomized experiment showed at a 95% confidence level that all the factors affected the product's output. About 91% yield was obtained and operating parameters were optimized at a temperature of around 61%. Methanol to oil ratio of 22.13, and catalyst loading of around 3.7 wt%. This chapter provided a step-by-step guide on how to carry out this experiment using design expert 13, a reduced Quadratic model with a significant P-value of 0.0325 shows the model is significant, as indicated by an f-value of 3.57. An F-value might be caused by noise only in 3.25% of cases. The run was reduced to 18 compared to the 20 runs originally used by Tshizanga et al.

Keywords: response surface methodology (RSM), central composite design (CCD), design of experiment (DOE), design expert

1. Introduction

It has always proven difficult to quickly select an appropriate experimental design, which can simply explicate many response factors. This sometimes leads to a quadratic surface model. CCD can be a choice for this kind of model. An experimental design called the central composite design (CCD) concept has emerged and has been very handy as part of the optimization process and search for the ideal product from ongoing batches. In statistics, a central composite design is an experimental kind of design, helpful in response surface methodology, for creating a second-order (quadratic) model for the response variables without having to use a complete three-level factorial experiment [1]. After performing the designed experiment, linear regression is deployed, sometimes iteratively, to obtain results. Coded Variables are frequently utilized when creating this design Most optimizations are done by screening all the potential variables [2]. Here, all the possible independent factors are first identified, and these factors are further improved before response surface methodology can finally be used to establish relationships between one or more process variables and their responses. The Central composite design is sometimes referred to as Box-Wilson central composite design and it has been chosen among researchers due to its accuracy.

2. Key terms in central composite design

Some important keywords will be mentioned throughout this chapter. This is to equip the readers with the terminology to understand fully the concept of Response surface methodology.

Response surface: These are the related variables. It involves a two or three-dimensional plot of the results of experimental data. Response surface methodology (RSM) is used to describe the use of experimental designs that give response surfaces from which information about the experimental system is deduced [3].

Factor: This can also be called the parameter or predictor. It is an entity that controls an outcome. The output Change is brought about by the manipulation or tweak of the input factor (s). They can be set and reset at different levels depending on the needs and conditions that affect the experiment.

Levels of the factors: The Design of experiments is named by the number of levels chosen for a factor, it could be a two or three-level design. It signifies the value of a factor that is prescribed in an experimental design. Levels could be high, mid, or low (three-level design) and only high and low (two-level design) is often coded as +1(high), 0(mid), and − 1(low). Selecting levels for an experiment often requires field experience. For example, for a three-level experiment, selecting the levels in a reactor would require some previous experience to decide 30°C(−1), 40°C (0) and 50°C (1) are suitable for low-level, mid-level, and high level respectively.

Blocking: This tool is used to eradicate the effects of external disturbances and in the process improve the efficiency of experimental design. External disturbances cause different forms of variations. The main goal is to arrange similar experiments runs into one group, so that the whole group becomes a homogeneous unit. For example in the transesterification reaction, A researcher is attempting to increase the yield of Biodiesel through Mean Absolute Errors (MAE). factors were considered for the initial experiment trials, which might have some impact on the yield. It is decided to study each factor in a two-level setting (i.e. a low value and a high value). Six experimental trials are chosen by the experimenter, but only four trials are possible to run per day. Here, each day can be handled separately as a different block [4].

Response: This is the result of the effect of an experiment, which is observed on account of changing the values of the predictors. For example the Yield, Selectivity, or Conversion of a reactant in a reactor.

Design of experiment (DOE): This is a statistical approach that involves planning, analyzing, conducting, and interpreting data obtained from experiments [3].

Randomization: While designing and running an experiment, there are several factors in the form of external disturbances often known as noise factors, which may influence how the experiment turns out. For example, variations in the quality of the raw material due to seasonal change, variations in the temperature, and their effects on the overall reaction yield may affect the result and such factors are difficult to control. Randomization is one of the methods to remove or reduce such errors occurring due to uncontrollable factors. Randomization helps in calculating the cumulative impact of the external disturbances if present in the process [3].

Model: This is an equation expressing the relationship between responses and the factors under study or investigation. Here the outcome can be denoted as a function of the experimental factors. For example, a model that has only one parameter x could be expressed as;

$$y = f(x) + \varepsilon \tag{1}$$

For two parameters model, it could represent as;

$$y = f(x_1, x_2) + \varepsilon \tag{2}$$

For the n parameters model, consider the following equation;

$$y = f(x_1, x_2 \dots \dots x_n) + \varepsilon \tag{3}$$

The function, $f(x)$ denotes the relationship between the parameters and the response (y) with the residuals (ε) and is depicted through a polynomial equation. Three different models are described:

Linear model: This is the simplest polynomial model that contains only linear terms and describes only the linear relationships between the variables and the responses. A linear model with two factors x_1, x_2 are expressed as:

$$y = b_0 + b_1x_1 + b_2x_2 + \varepsilon \tag{4}$$

Or can generally be represented as;

$$y = b_0 + \sum_{i=1}^{k} bixi + \varepsilon \tag{5}$$

Here, y is the outcome, b_i is the model coefficients, b_0 is the model intercept, i is the factor number from i to k, and x_i is the independent variables.

Interaction model: The interaction model holds some extra terms that depict interactions between various variables if any. For a two-factor, It is denoted as;

$$y = b_0 + b_1x_1 + b_2\,x_2 + b_{12}\,x_1x_2 + \varepsilon \tag{6}$$

Or generally as;

$$y = b_0 + \sum_{i=1}^{k} bixi + \sum_{i=1}^{k=1} \sum_{j=i+1}^{k} bijxixj + \varepsilon \tag{7}$$

b_o, b_i, and b_{ij} are the regression or the model coefficients for intercept, linear, and interaction terms, respectively, and x_i, and x_j are reaction factors.

Quadratic model: Quadratic terms are introduced in the model to help ascertain the optimal value. It helps to identify curvature that exists in the model. This model for two factors and interaction can be represented below:

$$y = b_0 + b_1x_1 + b_2x_2 + b_{12}x_1x_2 + b_{11}x_1^2 + b_{22}x_2^2 + \varepsilon \quad (8)$$

or generally a;

$$y = b_0 + \sum_{i=1}^{k} bixi + \sum_{i=1}^{k} biixi + \sum_{i=1}^{k=1} \sum_{j=i+1}^{k} bijxixj + \varepsilon \quad (9)$$

b_o, b_i, b_{ii}, and b_{ij} are the model coefficient for intercept, linear, quadratic, and interaction terms, respectively, and x_i, and x_j variables.

Note: The Symbol ε, in the model for eqs. (1) to (9) represents the residuals and the linear and interaction models are used during the screening stage.

Effects: This is often regarded as the coefficient of the variables, it can be distinguished from the *main effects;* which involve the factor's coefficient in the first-order model. *Interaction effect*; It is the coefficient of the products of linear terms. *Quadratic effect;* which denotes the coefficient of the square of the linear terms.

Replication: Replication means repeating the entire experiment or a part of it, under different operating conditions. It helps to obtain a projection of the experimental errors and to understand and estimate more specifically the factors and their interaction.

3. Response surface methodology for optimization design

The primary goal 0f optimization design is to minimize unfavorable or undesired outputs or maximize the desired outputs. Sometimes, simple linear and interaction models are not enough to provide a brilliant picture of the process. For this study, our goal is to increase the Biodiesel Yield from the transesterification of methanol and vegetable oil using a catalyst derived from the eggshell. The experiment has already been done and data is provided in this reference [4]. We will be using the information from this work to provide a thorough examination of central composite design in process optimization. The variables are reaction temperature, methanol-to-oil ratio, and catalyst weight. If these entities are positioned inside the region in which the experiment is to be conducted, we need a mathematical model that can represent curvature so that it has a local optimum. The best model is the quadratic model as shown in (eq. 9), which contains linear terms for all factors, squared terms for all factors, and products of all pairs of variables. Response surface designs are generally used for fitting quadratic models. A full factorial design with three levels for each input variable is one such design. Due to the excessive number of runs, that is not necessary to fit the model. It is typically not a good design. The CCD and Box–Behnken designs are the two most common designs generally used in response surface modeling although only central composite designs are explored in detail. In these types of designs, the variables take on three or five distinct levels, but not all combinations of these values appear in the design. The steps in CCD for Optimization are outlined below:

Preliminary stage: Here, the following steps are done:

- Choosing the factors and desired levels
- Determination of the Counts of experimental runs

- Calculation of alpha (α) and the axial values
- Selecting the response variables
- Carrying out the experiments
- Model selection

Analysis stage: At this stage, the following are done:

- ANOVA is conducted where the F-test and Lack of fitness are used to test for significance, Adjusted and predicted R^2 are also determined at this stage.
- Next, the model equation is built
- Comparing values predicted (from the model) and actual values
- Using 2D and 3D contour plots or graphs to visualize the response(s).

Decision-making stage: Here, the predicted and actual values are compared to determine the residuals using some useful parameters such as Adjusted R-Square, Mean Absolute Error (MAE), or Mean Square Error (MSE) is employed to assess the model performance and if okay with the result we can proceed to the final stage but if not we go back to the preliminary stage to see how we can adjust the model.

Optimization stage: At this stage, the model is ready to be deployed for the optimization process, design expert version 13 is very handy for this entire process, all we need is to specify the required values. The detail on how to determine the CCD components will be done later in this chapter.

4. Box: Behnken design (BBD)

The box design can adapt to the response surface full quadratic model [5]. BBD has no incorporated factorial or fractional factor designs, such as CC. In this design, the treatment combinations are at the midpoints of the edges of the cube and the center as shown in **Figure 1**. BBD is a rotatable design and needs three levels for each factor. BBD should be considered for experiments with more than two factors, and when it is expected that the optimum is known to lie in the middle of the factor ranges. A, B, and C represent factors A, B, and C respectively.

5. Central composite design

Central composite design (CCD): This is a unique kind of response surface design that can fit a full quadratic model. It is comprised of factorial also known as fractional factorial design with a center point attached to a group of stars or axial points. Using the included axial points is an effective method for calculating the coefficients of a second-degree polynomial for the factors [6]. A CCD can be denoted as a square (for two factors design) or a cube (for a three factors design) having corners, which represent the levels (high and low represented as +1, −1 respectively), a star or axial

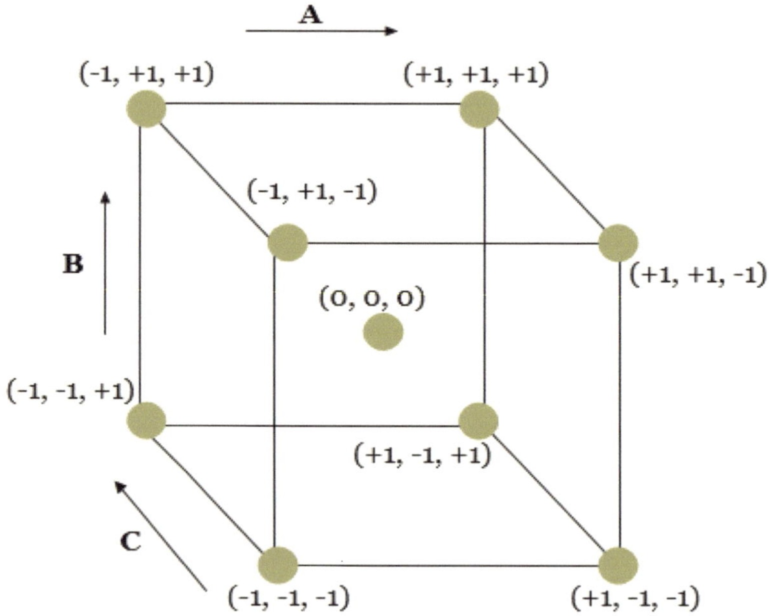

Figure 1.
A representation of box–Behnken design.

points along the axes at or outside the square helps to account for the curvature and a center point at the origin. The general model for a two-factor full factorial CCD is represented graphically in **Figure 2** below.

Figure 3 displays a three-factor lay out for a CCD made up of a full factors factorial that forms the cube where each side is coded −1 and + 1 just like in **Figure 2** above. The Stars stand for axial points and alpha is the distance from the edge of the cube to the stars.

6. Types of central composite design

There are three types of CCD namely:

- Circumscribed Central Composite Design (CCC)
- Inscribed Central Composite Design (CCI)
- Face-Centered Central Composite Design (CCF)

The CCC is a type of CCD in which the location of the axial points forms new extremes from the already attained levels of the factorial factors. The new extremes are determined by a value called alpha (distance between the new extreme and the edges of the factorial points) making it up to 5 levels. It is often determined to achieve a *rotatable* design [7].

The CCI type is a modified form of CCC. The axial points are scaled to be within the limits of the factorial factor [8]. The CCI is also a rotatable type and has 5 levels just like the CCC.

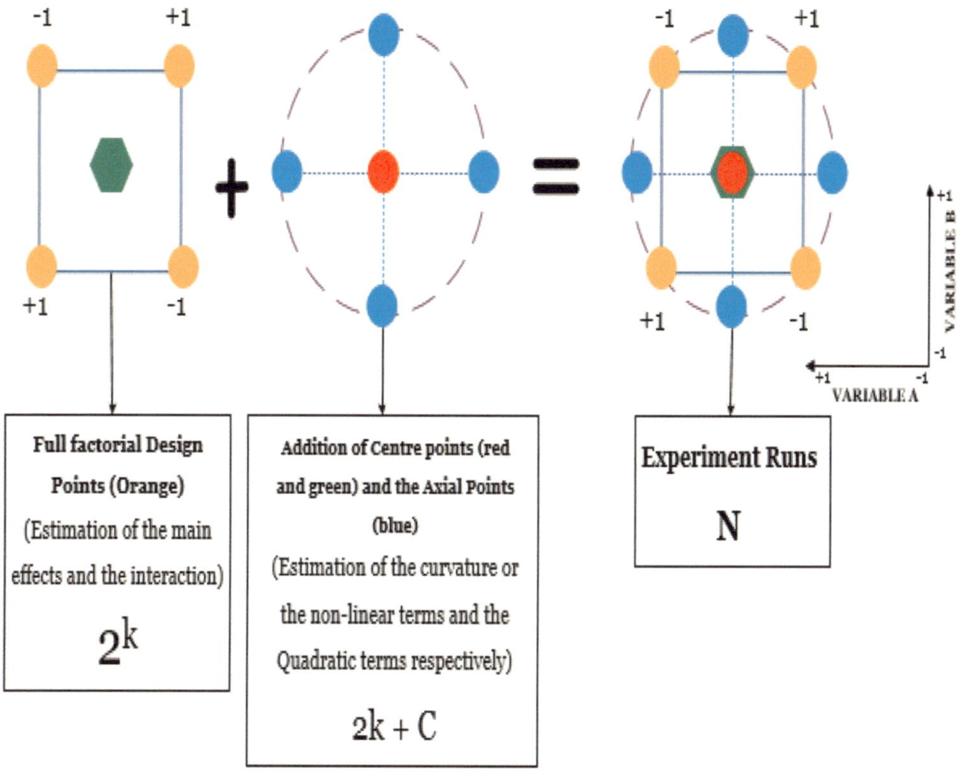

Figure 2.
A visual depiction of the CCD model for determination of total runs for all experiments for two factors full factorial design. K in the model is the number of factors, C is the replicated central points that help to eliminate pure error and N is the experiment runs required for the design.

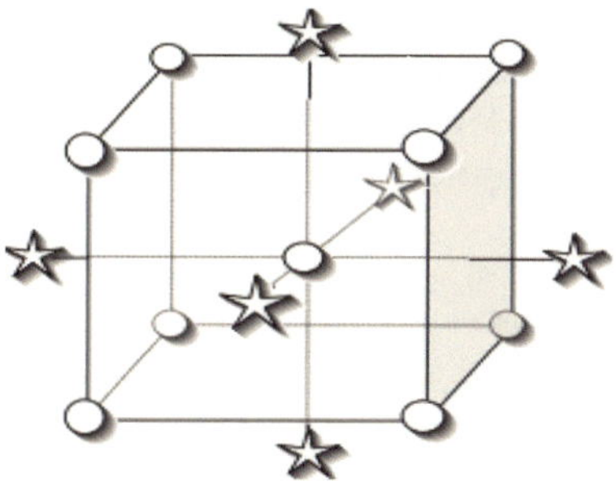

Figure 3.
A graphical representation of three factors in a full factorial design.

For the CCF, the axial points correspond to the center points for each side of the cube in **Figure 3** above (three factors designs) and they are non-rotatable [9]. It has only 3 levels. **Figure 4** below provides more insight into this type of CCD.

7. Determining the components of central composite design

Before starting the CCD optimization process, we need to provide a walk-through on how to calculate all the required parameters to build the model.

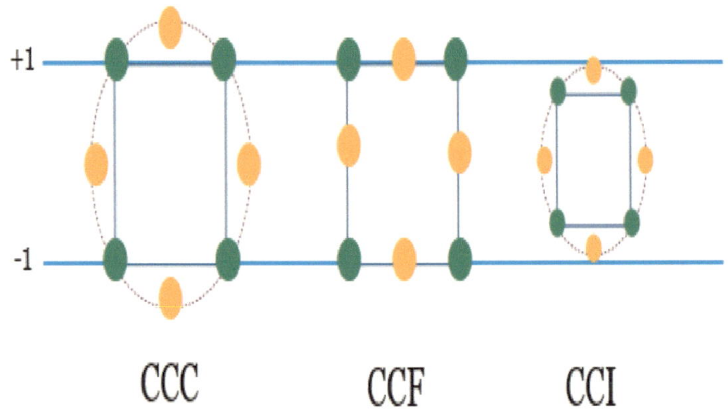

Figure 4.
Three types of central composite Design6.

8. Calculating the number of experiment runs

To design a CCD experiment for two levels (+1 and − 1 levels of factors) full factorial design is represented by 2^k, then the axial points as represented by **Figure 2** are given as 2 k, let C represent the center points and n, the number of times the experiment is repeated to eliminate errors. Then the total number of experiment runs is given as:

$$N = 2^k + 2k + nC \tag{10}$$

Where k is the number of factors selected for the experiments. For our case, we have three (3) factors, i.e. Temperature, methanol-ratio, catalyst weight, and 4 repetitions. By substituting k = 3, C = 1, and n = 4 (i.e. 4 repetitions), then N = 18 runs. Luckily, design expert 13 will automatically generate this value once the number of factors and repetitions are provided. Keep in mind the number of center points can also be adjusted by clicking the options button on the software, in this case, we will just use.

9. Calculating alpha (α)

As can be seen that immediately after the factors, n, and C are provided the alpha is automatically calculated, this is because the minimum parameters to calculate it has been specified, now it will be shown how the program generates this value. As has been discussed earlier Alpha is the distance between the new extreme axial points and the edge formed from the factorial levels. Now the following equation will calculate this **α** value for any factor.

$$\alpha = \left(2^k\right)^{1/4} \tag{11}$$

For our case, k is 3, and therefore **α** = **1.68179** which is in line with the value created by the software. Consider the **Table 1** below for k from 2 to 5 factors and their corresponding values.

Factor (k)	Alpha (α)
2	1.41421
3	1.68179
4	2
5	2.37841

Table 1.
Factors and corresponding α values.

10. Calculating axial values

Before determining the axial points, the table below shows the factors levels and center points that will be used to compute the axial points. The center points are coded as 0, while low and high levels are designated as −1 and + 1 respectively. Also keep in mind that the experiment has already been performed and data provided from the work of (**Table 2**) Tshizanga et al., [4].

To compute the Axial values, the first thing to do is to find the α that can be added or subtracted from the factor levels (low and high) and the center points. Adding α to factor levels can be coded as + α (higher axial value) while subtracting α from the mean factor levels is however coded as – α (lower axial value). These additional two coded values (+α and – α) are axial and they make the factors up a total of 5. The two equations are given below:

$$+\alpha = X + \left(\alpha \text{ x } \frac{High\ level - Low\ level}{2}\right) \tag{12}$$

$$\alpha = X - \left(\alpha \text{ x } \frac{High\ level - Low\ level}{2}\right) \tag{13}$$

Where α can be found using (eq. 11) although calculated as **1.68179** and X is given by:

$$(\text{Low level} + \text{centre point} + \text{High level})/\text{k} \tag{14}$$

And k is the number of variables, in this case, k is 3. At this point let us get our hands dirty with calculating the values for these 3 factors.

For Temperature:

$$\boldsymbol{X_1 = \frac{60 + 65 + 70}{3} = 65}$$

Factors	Symbols	Low level (−1)	Centre point (0)	High level (+1)
Temperature (°C)	X_1	60	65	70
Methanol to Oil ratio	X_2	15:1	22.5:1	30:1
Catalyst Weight (wt%)	X_3	2.0	3.5	5

Table 2.
Experimental ranges of the independent variable.

$$(app.\ To\ 4\ d.p) \qquad +\alpha = 65 + \left(1.68179\ \text{x}\frac{70-60}{2}\right) = 73.4090°C$$

$$-\alpha = 65\text{–}\left(1.68179\ \text{x}\frac{70-60}{2}\right) = 56.5911°C$$

For Methanol-Oil ratio:

$$X_2 = \frac{15+22.5+30}{3} = 22.5\ (22.5:1)$$

$$+\alpha = 22.5 + \left(1.68179\ \text{x}\frac{30-15}{2}\right) = 35.1134\ (35.1134:1)$$

$$-\alpha = 22.5\text{–}\left(1.68179\ \text{x}\frac{30-15}{2}\right) = 9.8866\ (9.8866:1)$$

For Catalyst weight:

$$X_3 = \frac{2+3.5+5}{3} = 3.5$$

$$+\alpha = 3.5 + \left(1.68179\ \text{x}\frac{5-2}{2}\right) = 6.0227\ \text{wt\%}$$

$$-\alpha = 3.5\text{–}\left(1.68179\ \text{x}\frac{5-2}{2}\right) = 0.9773\ \text{wt\%}$$

Currently, we have succeeded and step by step discussed how the software generated the alpha (α) and axial values as the components of the CCD, below is a table including these axial points (**Table 3**).

Upon specifying the required parameters for the CCD model the software will generate a table where the experiment will now be conducted to enable determining the response for each experiment run. For this case study, our response is biodiesel yield which can be determined from methyl ester and waste vegetable oil weight using the following equation:

$$\text{Yield}\ (\%) = \frac{Weight(Biodiesel)}{Weight(Oil)}\ \text{x}\ 100 \qquad (15)$$

Table 4 factors' coded values organized in the standard order.

Factors	Symbols	Lower axial point (−α)	Low level (−1)	Centre point (0)	High level (+1)	Higher axial point (+α)
Temperature (°C)	X_1	56.5911	60	65	70	73.4090
Methanol to Oil ratio	X_2	9.8866	15:1	22.5:1	30:1	35.1134
Catalyst Weight (wt%)	X_3	0.9773	2.0	3.5	5	6.0227

Table 3.
Experimental ranges of independent variables including calculated axial (star) values.

Std	Run	Factor 1 A: Temperature (0C)	Factor 2 B:Methano-Oil r ...	Factor 3 C:Catalyst Weight wt%	Response 1 Biodiesel Yield %
1	3	−1.000	−1.000	−1.000	
2	15	1.000	−1.000	−1.000	
3	11	−1.000	1.000	−1.000	
4	4	1.000	1.000	−1.000	
5	2	−1.000	−1.000	1.000	
6	12	1.000	−1.000	1.000	
7	17	−1.000	1.000	1.000	
8	8	1.000	1.000	1.000	
9	7	−1.682	0.000	0.000	
10	1	1.682	0.000	0.000	
11	16	0.000	−1.682	0.000	
12	14	0.000	1.682	0.000	
13	13	0.000	0.000	−1.682	
14	6	0.000	0.000	1.682	
15	9	0.000	0.000	0.000	
16	18	0.000	0.000	0.000	
17	5	0.000	0.000	0.000	
18	10	0.000	0.000	0.000	

Table 4.
Factors' coded values organized in the standard order.

Immediately after we fill up the required CCD components, the design expert will provide a table for coded levels of factors. This is being used to use as a guide to specifying the actual values and their corresponding responses. The experiment will be repeated four (4) times instead of 6 times (as done by the original researchers) to reduce the experiment runs. The two results will be compared after the optimization stage. **Table 4** shows the coded factors, and **Table 5** shows the actual values and their responses after experimenting in the laboratory.

At this point, we can now replace the coded values with the actual values from the previous calculations. The factor columns were generated with a particular pattern, but it's beyond the scope of this chapter, to learn more about this we recommend reading "RSM simplified by Anderson and Whitcomb" [10].

11. Results and analysis

Note: When entering the values for the methanol to oil ratio in design expert, you can ignore all the 1's, since the value of Oil concerning Methanol is always a unit for all the experiment runs.

We can now delve into understanding the data collected to build the model, perform analysis, and finally carry out the optimization. All these steps will be done in design expert software.

Std	Run	Factor 1 A: Temperature (0C)	Factor 2 B:Methanol-Oil ratio	Factor 3 C:Catalyst Weight wt%	Response 1 Biodiesel Yield %
1	3	60	15	2	69.29
2	15	70	15	2	35.2
3	11	60	30	2	73.39
4	4	70	30	2	10.66
5	2	60	15	5	61.29
6	12	70	15	5	50.23
7	17	60	30	5	71.48
8	8	70	30	5	22.06
9	7	56.591	22.5	3.5	62.18
10	1	73.409	22.5	3.5	79
11	16	65	9.88655	3.5	42.68
12	14	65	35.1134	3.5	39.5
13	13	65	22.5	0.977311	66.52
14	6	65	22.5	6.02259	74.04
15	9	65	22.5	3.5	80.46
16	18	65	22.5	3.5	89.35
17	5	65	22.5	3.5	90.98
18	10	65	22.5	3.5	89.52

Table 5.
Actual factors' values arranged in the standard order after the experiment.

12. Understanding the data

The reason for this is basically to understand the relationship that exists in the data. In a more statistical sense, we need to know if there is a strong correlation between the variables and the response. If to some extent there exists an intra-correlation among the factors then one of them has to be removed because it will eventually harm the model. Design expert has provided a wonderful dashboard where we can carefully learn more about the data we have collected and make some sense of it. At the left of the software, we will see the *information* part of it. The summary, graphs columns, and evaluation subsections are the places to dig the nuggets from the data.

In the *summary section* we will see the summary statistics of the data, i.e. the number of experiment runs, type of designs and model, minimum, maximum, mean, standard deviations of the responses, and Ratio of maximum to minimum response values (**Table 6**).

We have seen that the mean response is quite far away from the minimum and maximum response, this is the primary reason for building this model to test the statistical significance of this result. If we are okay with the significance we can go ahead with the model built in the evaluation tab (**Table 7**).

Factor	Name	Units	Type	SubType	Minimum	Maximum	Coded Low	Coded High	Mean	Std. Dev.
A	Temperature	(0C)	Numeric	Continuous	56.59	73.41	−1 ↔ 60.00	+1 ↔ 70.00	65.00	4.24
B	Methanol-Oil ratio		Numeric	Continuous	9.89	35.11	−1 ↔ 15.00	+1 ↔ 30.00	22.50	6.36
C	Catalyst Weight	wt%	Numeric	Continuous	0.9773	6.02	−1 ↔ 2.00	+1 ↔ 5.00	3.50	1.27

Table 6.
Summary statistics of the factors.

Response	Name	Units	Observations	Minimum	Maximum	Mean	Std. Dev.	Ratio
R1	Biodiesel Yield	%	18.00	10.66	90.98	61.55	23.49	3.53

Table 7.
Summary statistics of the response.

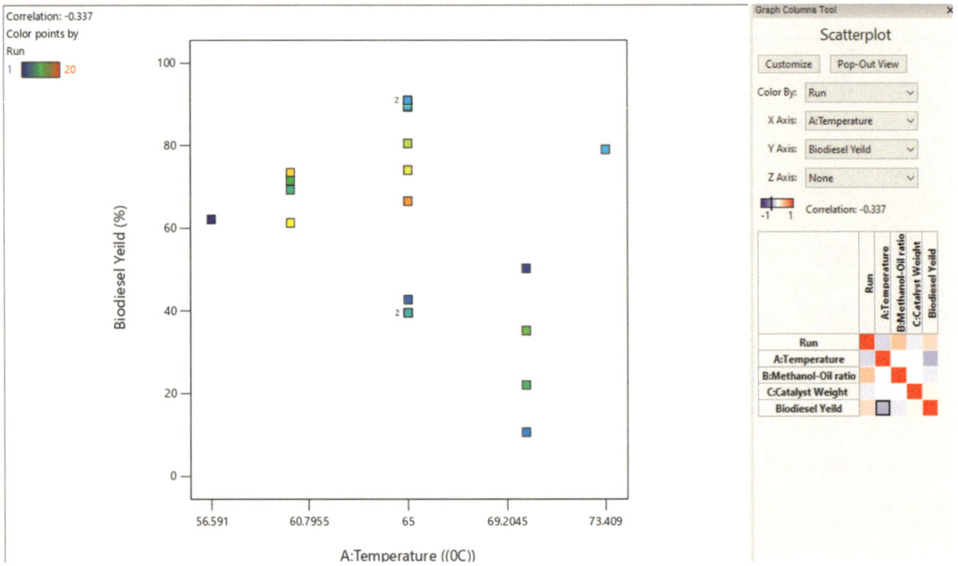

Figure 5.
A scatterplot of temperature vs. biodiesel yield.

Moving over to the *graphs column* section, there are scatter plots, histograms, and Box-plots. To make the most sense of this data, the scatter plot is most handy since it tells how the factors are correlated to each other, the drop-down at the top left helps to select the factors to show scatter plots or the correlations plots at the bottom to help display correlations as values between −1 and 1 (blue to red). Values close to −1 show a strong negative correlation and values close to +1 show a strong positive correlation. Now we will go ahead and display the scatter plots of each factor and the response under the factor that mostly affects the biodiesel yield in **Figures 5–7** respectively.

The plots have shown that temperature mostly negatively affects yield. The correlation plot in **Figure 7** confirmed this claim since the box with the most blueish color lies between the temperature and Biodiesel columns. As described in the correlation plot in **Figure 8** from design expert software.

Finally, this section provides a unique tab called ***evaluation***, where the model name is selected and all their parameters are shown. In this case, a quadratic model has been selected by the software which is the best for CCD. There are two tabs i.e. the results and graphs where the model parameters are evaluated.

In the model tab, the model terms are related to the Significant factor, Variance Inflation Factor, R-Squared, and Power of the model as shown in the table below.

12.1 Model terms

Power calculations are performed using response type "Continuous" and parameters:

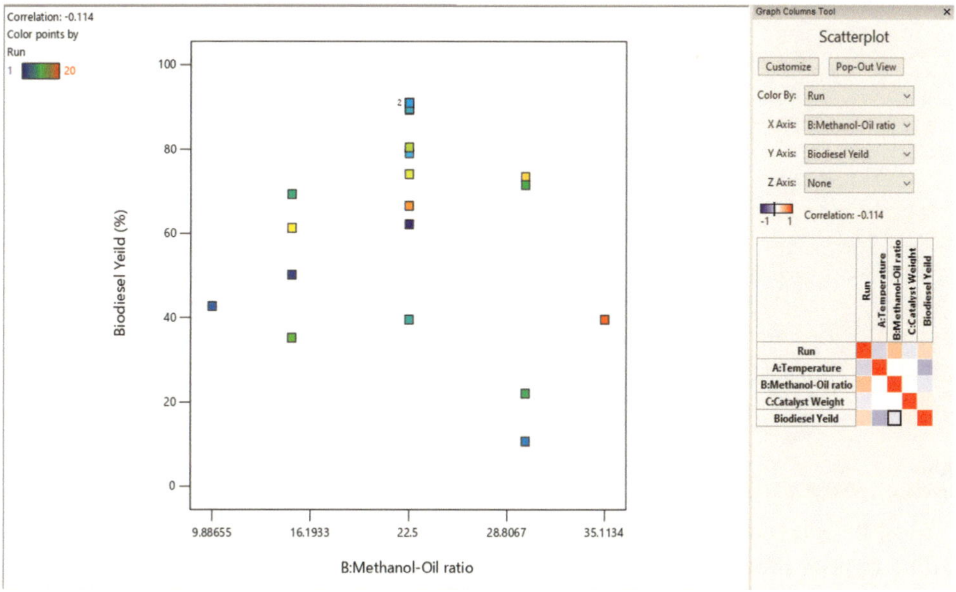

Figure 6.
A scatterplots of methanol/oil ratio against biodiesel yield.

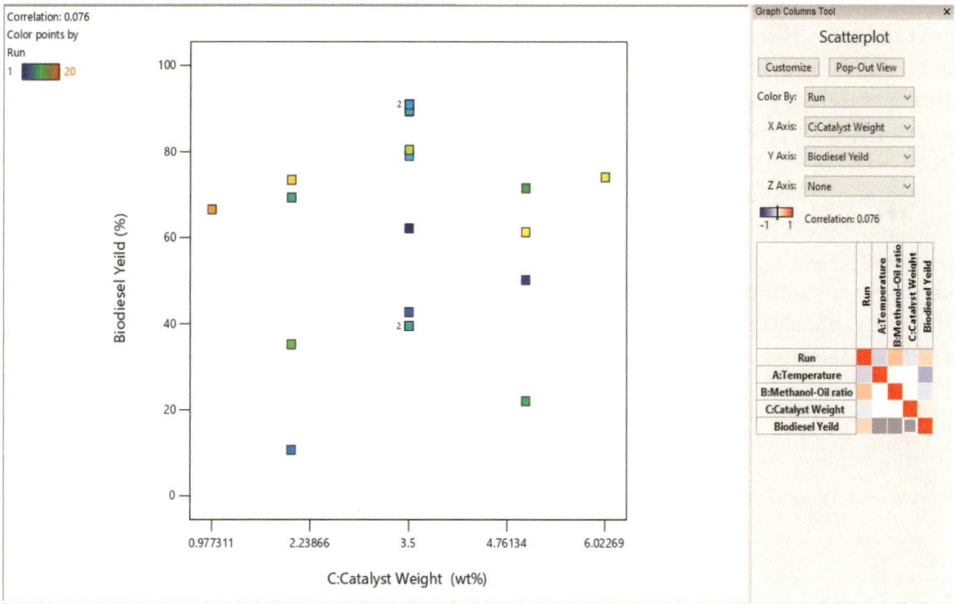

Figure 7.
A scatterplots of catalyst weight against biodiesel yield.

Delta = 2, Sigma = 1.

Power is evaluated over the −1 to +1 coded factor space. Standard errors should be similar to each other in a balanced design. Lower standard errors are better.

The ideal VIF value is 1.0. VIFs above 10 should cause concern. VIFs above 100 should cause alarm, indicating coefficients are poorly estimated due to multicollinearity.

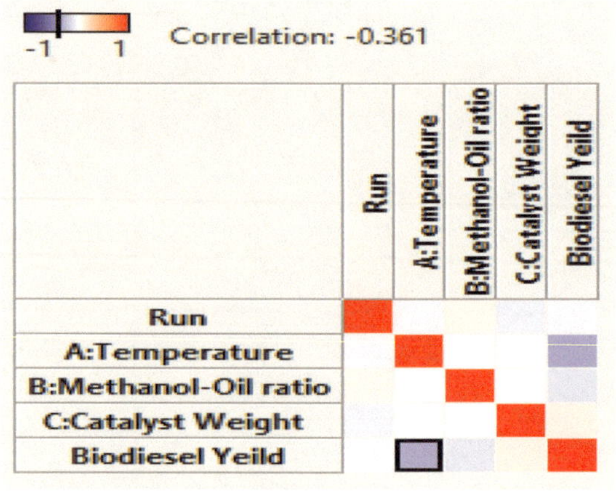

Figure 8.
Correlation plots of all the relationships that exist in the data.

Ideal R_i^2 is 0.0. High R_i^2 means terms are correlated with each other, possibly leading to poor models. If the design has multilinear constraints, then multicollinearity will exist to a higher extent. This inflates the VIFs and the R_i^2, rendering these statistics useless. Use FDS instead.

The Power Calculation is the estimated chance to find a significant effect out of the current evaluation model. Power depends on the size and structure of the design, the signal-to-noise ratio (number of standard deviations) for the effect, and the model evaluated. The Options button on the Model tab allows the user to define three signal-to-noise ratios that define the number of standard deviations to use. If the power is not large enough (80% or more) for a reasonably sized effect, then the design is underpowered. As can be seen in **Table 8**, we may consider removing the interaction terms since they have lower power. This will be done after analyzing the model and the p-value is higher than 0.05. This means they have affected the performance of the model. Power is an inappropriate tool to evaluate response surface designs. Use prediction-based metrics provided in this program via Fraction of Design Space (FDS) statistics. Click on the Graphs tab to find the FDS graph. More information about FDS is available in the Help. Be sure that the model you selected contains only terms you expect to be significant (**Table 9**).

Term	Standard Error*	VIF	R_i^2	Power
A	0.2706	1	0.0000	91.4%
B	0.2706	1	0.0000	91.4%
C	0.2706	1	0.0000	91.4%
AB	0.3536	1	0.0000	72.2%
AC	0.3536	1	0.0000	72.2.%
BC	0.3536	1	0.0000	72.2%
A^2	0.2634	1.01827	0.0179	99.9%
B^2	0.2634	1.01827	0.0179	99.9%
C^2	0.2634	1.01827	0.0179	99.9%

Table 8.
Model parameters.

Run	Leverage	Space Type
1	0.6073	Axial
2	0.6698	Factorial
3	0.6073	Axial
4	0.6698	Factorial
5	0.1663	Center
6	0.6073	Axial
7	0.1663	Center
8	0.1663	Center
9	0.6698	Factorial
10	0.6698	Factorial
11	0.6698	Factorial
12	0.6693	Factorial
13	0.1663	Center
14	0.1663	Center
15	0.6073	Axial
16	0.6698	Factorial
17	0.6698	Factorial
18	0.1663	Center
19	0.6073	Axial
20	0.6073	Axial
Average	0.5000	

Table 9.
Leverage.

12.2 Leverage

The leverage data as shown in the table above is the potential for a design point to influence the fit of the model coefficients, based on its position in the design space. Leverages approaching or at 1 indicate that point will influence the model. A leverage of 1 means the model must exactly fit the observed value. A good design avoids leverages approaching 1. A design for the same model but having more runs will tend to have a lower leverage for each point.

Watch for leverages close to 1.0. Consider replicating these points or make sure they are run very carefully.

The Graphs tab contains the FDS, Perturbation, interactions, Contour, Cube, and 3D Surface Plots to help understand the data and the model parameters.

12.3 FDS graph

The FDS graph is used to compute the volume of the design space that has predicted variance less than or equal to the specified value. The fraction of the design space is calculated as this volume divided by the entire volume of the design space.

The goal is to make a single plot that shows the cumulative fraction of the design space on the x-axis (from zero to one) versus the prediction variance on the y-axis.

For exploration and optimization, we advise an FDS score of at least 0.8, or 80%, and for stability and robustness testing, such as showcasing the design space for quality by design (QbD) work, 100%. Options for assessing the FDS in relationship to four different error categories are provided by the FDS Graph tool, i.e. Mean, prediction intervals, Difference between pairs of Observations, and Tolerance. We are using the Mean error type since the aim of this experiment is to find the optimized factor settings for specific response goals. **Figure 9** below is the visualization for the FDS graph.

There are three parameters: **delta, sigma, and alpha** for each type of error and a fourth parameter is **Proportion** for the Tolerance type of error.

delta specifies the maximum acceptable half-width (margin of error) of the respective interval for the Mean, Pred, and Tolerance error types. One best way to find the delta is to answer the question, “plus or minus how much is an acceptable estimate?”

sigma is an estimate for the standard deviation that will appear on the ANOVA. It can be obtained from previous work with this system, work from a similar system, or outright guessing. A smaller sigma can be entered to enhance the FDS if the unexplained nuisance fluctuation can be reduced during the experiment.

Alpha is the used significance level throughout the statistical analysis. Our default is 0.05 or 5%. It is a type I error acceptable risk. FDS rises as alpha increases. The critical value is calculated using alpha/2 for two-sided intervals and alpha for one-sided intervals.

Proportion is only used for the Tolerance type of error. It is the percentage of the individual outcomes that must fall within the tolerance range. Building a larger design and raising the delta will boost the FDS score, reducing the sigma, increasing the alpha, and/or decreasing the Proportion [11, 12].

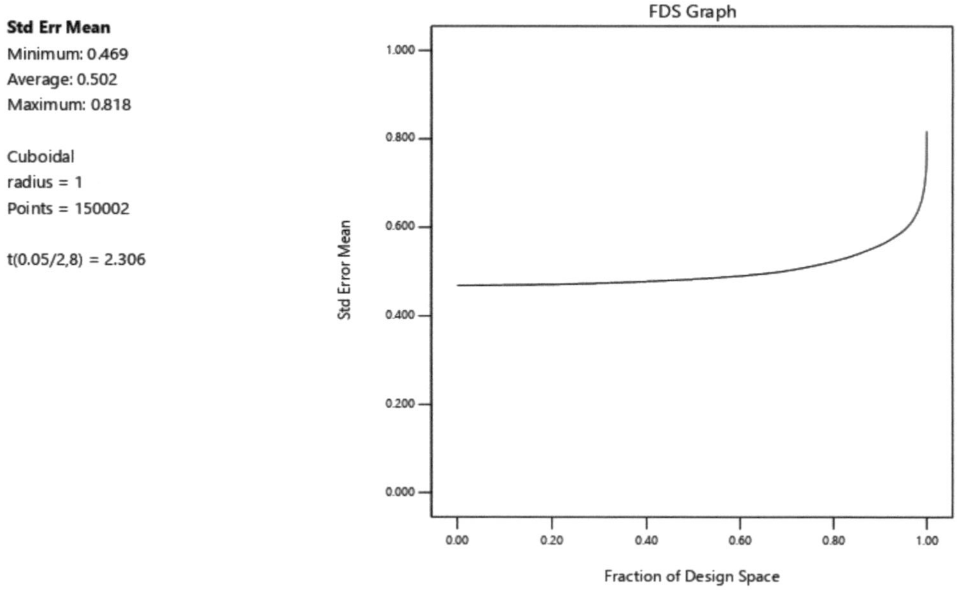

Figure 9.
FDS graph.

12.4 Interaction

When the reaction varies depending on the settings of two elements, there has been an interaction. They will display two non-parallel lines, showing that one element has an impact on them and depends on the level of the other. **Figure 10** displays the standard error of the design with interactions of the model parameters.

13. Analysis

In the Analysis Section in design expert, select no transform in the configure tab and start the analysis using the button at the button. The interface should appear like **Figure 11** below.

You can take advantage of the advanced options button to customize the model like changing from coded to actual factors for factors coding (It is not recommended though).

13.1 Fit summary

The regression calculations to fit all of the polynomial models to the chosen answer are started when the Fit Summary button is clicked. All model terms' effects are calculated by the program. It produces statistics such as p-values, lack of fit, and R-squared values for comparing the models. The fit summary output is shown on screen in a report which can also be printed and/or copied to another application detected, The "Suggested" model will be highlighted and noted by the program. On the Model panel, this is set as the default model. We Look for the following (**Table 10**):

- A high-order model explains significantly more of the variation that is in the response (p-value small).

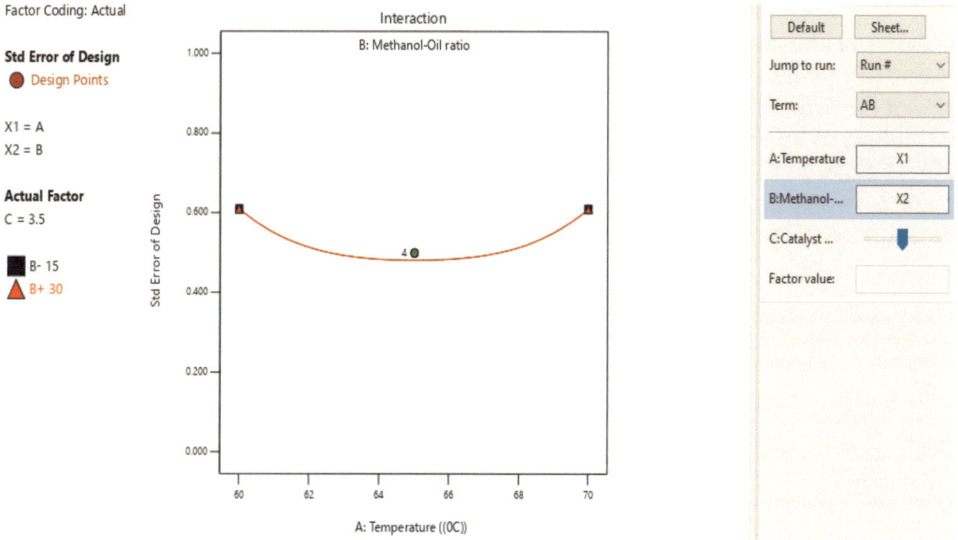

Figure 10.
Std error of the design with interactions of the model parameters.

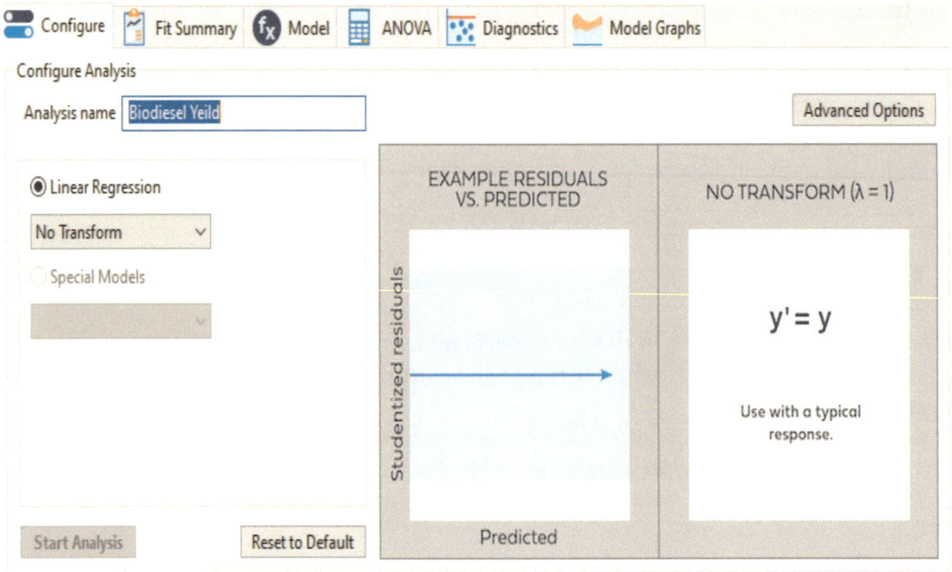

Figure 11.
Starting the analysis.

- Insignificant lack of fit (p-value >0.10).
- Adjusted R-squared and predicted R-squared have a reasonable level of agreement (within 0.2 of each other).

Note: Aliased Models should entirely be avoided.

13.2 Sequential model sum of squares

Table 11 shows the sum of squares, degree of freedom, mean square, F-value, and p-value of the design model. The Sequential Model Sum of squares is the sum of the squared deviations from the mean for each model. The SS for the Mean is calculated first, followed by the Blocks (if applicable), Linear model, Quadratic model, Special Cubic, Cubic, Residuals, and Total.

For each source, the sum of squares divided by the degrees of freedom yields the mean square. This is used to compute the F-value for the models.

Response 1: Biodiesel Yield					
Source	**Sequential p-value**	**Lack of Fit p-value**	**Adjusted R^2**	**Predicted R^2**	
Linear	0.4974	0.0082	−0.0302	−0.3670	
2FI	0.7771	0.0060	−0.1914	−1.6146	
Quadratic	0.0279	0.0157	0.4440	−0.9458	Suggested
Cubic	0.0635	0.0354	0.8292	−6.2531	Aliased

Table 10.
Fit summary.

Source	Sum of Squares	df	Mean Square	F-value	p-value	
Mean vs. Total	68182.63	1	68182.63			
Linear vs. Mean	1421.30	3	473.77	0.8337	0.4974	
2FI vs. Linear	726.96	3	242.32	0.3687	0.7771	
Quadratic vs 2FI	4775.31	3	1591.77	5.19	0.0279	Suggested
Cubic vs. Quadratic	2076.83	4	519.21	5.51	0.0635	Aliased
Residual	376.84	4	94.21			
Total	77559.86	18	4308.88			

Table 11.
Modeling sequentially, sum of squares.

The F-value is used to test the significance of adding new model terms to those terms already in the model. For instance, the meaning of the linear terms remains tested after removing the effect of the average and the blocks. Then, the significance of the quadratic terms is tested after removing the average, block, and linear effects. And so on. Select the polynomial with the highest order and where the additional terms are significant and the model is not aliased.

13.3 Model summary statistics

R-squared is the correlation coefficient for the model. It should be close to one. We recommend using the Adjusted R-squared for DOE evaluation.

The amount of variation that can be explained by the model is shown by the adjusted R-squared. This is the R-squared value after adjusting for how many terms are in the model relative to the number of design points. The Model summary statistics is shown in **Table 12**.

Predicted R-Squared is calculated from the PRESS statistic, this represents the amount of variation in new data explained by the model. A negative Predicted R-squared means that the overall mean is a better predictor than this model.

Focus on the model maximizing the **Adjusted R^2** and the **Predicted R^2**.

13.4 Lack of fit tests

The data for the Lack of fit Test is shown above in **Table 13**. This is the p-value associated with the Lack of Fit calculation for this model. The best model should have an insignificant p-value. A typical cutoff would be a p-value >0.10 to conclude an insignificant lack of fit.

The selected model should have an insignificant lack of fit.

13.5 ANOVA for quadratic model

Table 14 is the Anova data which is used to test for the significance of the result obtained. Model Probability (a.k.a. p-value) is the probability that the model F statistic is at least the computed value even though the truth is there are no factor effects (the data produced false effects). Probabilities less than the acceptable risk (alpha, by default 0.05) are deemed significant and indicate that there is a model effect. Values greater than the alpha risk suggest no significant effect.

Source	Std. Dev.	R^2	Adjusted R^2	Predicted R^2	PRESS	
Linear	23.84	0.1516	−0.0302	−0.3670	12818.28	
2FI	25.64	0.2291	−0.1914	−1.6146	24517.49	
Quadratic	17.51	0.7383	0.4440	−0.9458	18246.34	Suggested
Cubic	9.71	0.9598	0.8292	−6.2531	68014.21	Aliased

Table 12.
Model summary statistics.

Source	Sum of Squares	df	Mean Square	F-value	p-value	
Linear	7886.78	11	716.98	31.11	0.0082	
2FI	7159.83	8	894.98	38.83	0.0060	
Quadratic	2384.52	5	476.90	20.69	0.0157	Suggested
Cubic	307.68	1	307.68	13.35	0.0354	Aliased
Pure Error	69.15	3	23.05			

Table 13.
Lack of fit tests.

The degree to which the model fits the data is measured by lack of fit. A strong lack of fit ($p < .05$) is an undesirable property because it shows that the model does not fit the data well. It is desirable to have little lack of fit ($P > 0.1$).

The model is not significant in comparison to the noise, according to the **model's F-value** of 2.51. The likelihood of noise causing an F-value this large is 10.49%.

Source	Sum of Squares	df	Mean Square	F-value	p-value	
Model	6923.57	9	769.29	2.51	0.1049	not significant
A-Temperature	1218.74	1	1218.74	3.97	0.0813	
B-Methanol-Oil ratio	140.27	1	140.27	0.4573	0.5179	
C-Catalyst Weight	62.29	1	62.29	0.2031	0.6642	
AB	561.12	1	561.12	1.83	0.2132	
AC	165.07	1	165.07	0.5382	0.4841	
BC	0.7564	1	0.7564	0.0025	0.9616	
A^2	836.53	1	836.53	2.73	0.1372	
B^2	4358.27	1	4358.27	14.21	0.0055	
C^2	859.23	1	859.23	2.80	0.1327	
Residual	2453.67	8	306.71			
Lack of Fit	2384.52	5	476.90	20.69	0.0157	significant
Pure Error	69.15	3	23.05			
Cor Total	9377.24	17				

Table 14.
Anova for quadratic model.

Fit Statistics			
Std. Dev.	**17.51**	**R^2**	**0.7383**
Mean	61.55	Adjusted R^2	0.4440
C.V. %	28.46	Predicted R^2	−0.9458
		Adeq Precision	4.8223

Table 15.
Fit statistics.

Model terms are considered significant when the P-**value** is less than 0.0500. B^2 is a significant model term in this situation. The model terms are not significant if the value is greater than 0.1000. Model reduction may enhance the model if it has a lot of unnecessary terms (except those needed to maintain hierarchy).

The significance of the lack of fit is indicated by the **lack of fit F-value** of 20.69. A significant Lack of Fit F-value can only be caused by noise in 1.57 percent of cases. A significant lack of fit is undesirable, we want the model to fit.

A negative Predicted R^2 as shown in the Fit Statistics data in **Table 15** implies that the overall mean may be a better predictor of your response than the current model. In some cases, a higher-order model might be more accurate.

Adeq Precision: The ratio of signal to noise is measured by Adeq Precision. A ratio of at least 4 is preferred. Your ratio of 4.822 shows an adequate signal. This model can be used to navigate the design space.

14. Decision

From the ANOVA result, it is obvious the model cannot be deployed like this, we need to tweak it a bit before using it for optimization, or else the solutions provided by it will be misleading. Now we will remove the interaction terms from the model since they have lower power (see **Table 8**). We will only repeat the ANOVA section after this change (**Table 16**).

14.1 ANOVA for reduced quadratic model

The model is significant, as indicated by the model's F-value of 3.57 The likelihood of noise producing an F-value this large is only 3.25.

Model terms are considered significant when the P-value is less than 0.0500. In this case, B^2 is a crucial model term in this instance. Model terms are not significant if the value is higher than 0.100. Model reduction may enhance your model if it has a large number of unnecessary terms (excluding those necessary to maintain hierarchy).

The Lack of Fit F-value of 16.87 implies the Lack of Fit is significant. There is only a 2.02% chance that a Lack of Fit F-value this large could occur due to noise. A significant lack of fit is not okay – we want the model to fit. But there is a little bit more improvement than before. So we can work with this model (**Table 17**).

14.2 Fit statistics for RQM

A negative **Predicted R^2** implies that the overall mean may be a better predictor of your response than the current model. In some cases, a higher-order model may also predict better.

Source	Sum of Squares	df	Mean Square	F-value	p-value	
Model	6196.61	6	1032.77	3.57	0.0325	significant
A-Temperature	1218.74	1	1218.74	4.21	0.0646	
B-Methanol-Oil ratio	140.27	1	140.27	0.4851	0.5006	
C-Catalyst Weight	62.29	1	62.29	0.2154	0.6516	
A^2	836.53	1	836.53	2.89	0.1170	
B^2	4358.27	1	4358.27	15.07	0.0026	
C^2	859.23	1	859.23	2.97	0.1127	
Residual	3180.62	11	289.15			
Lack of Fit	3111.47	8	388.93	16.87	0.0202	significant
Pure Error	69.15	3	23.05			
Cor Total	9377.24	17				

Table 16.
ANOVA for reduced quadratic model.

Adeq Precision measures the signal-to-noise ratio. A ratio greater than 4 is desirable. Our ratio of 5.459 indicates an adequate signal. This model can be used to navigate the design space. And there is an improvement in the Adjusted R^2 using this reduced Quadratic Model.

14.3 Coefficients in terms of coded factors

The coefficient estimate data in **Table 18** represents the estimated coefficient and shows the anticipated change in response for each unit change in the factor value. The intercept in an orthogonal design is the overall average response of all the runs. The coefficients are adjustments around that average based on the factor settings. The VIFs are 1 when the factors are orthogonal; Multi-collinearity is indicated by VIFs that are more than 1, and the higher the VIF, the more severe the correlation of components VIFs of fewer than 10 are generally acceptable.

14.4 Final equation in terms of coded factors

You can apply the equation in terms of coded factors in **Table 19** to make predictions about the response for given levels of each factor. By default, the factors' high levels are coded as +1 and their low levels as −1. By comparing the factor coefficients, the coded equation can be used to determine the relative importance of the elements.

Std. Dev.	17.00	R^2	0.6608
Mean	61.55	Adjusted R^2	0.4758
C.V. %		Predicted R^2	−0.3457
		Adeq Precision	5.4594

Table 17.
Fit statistics for RQM.

Factor	Coefficient Estimate	df	Standard Error	95% CI Low	95% CI High	VIF
Intercept	88.05	1	8.49	69.37	106.74	
A-Temperature	−9.45	1	4.60	−19.57	0.6808	1.0000
B-Methanol-Oil ratio	−3.20	1	4.60	−13.33	6.92	1.0000
C-Catalyst Weight	2.14	1	4.60	−7.99	12.26	1.0000
A^2	−8.13	1	4.78	−18.66	2.39	1.08
B^2	−18.56	1	4.78	−29.09	−8.04	1.08
C^2	−8.24	1	4.78	−18.76	2.28	1.08

Table 18.
Coefficients as codified factors.

14.5 Final equation in using actual factors

The equation in terms of actual factors in **Table 20** can be used to make predictions about the response for given levels of each factor. Here, the levels should be specified in the original units for each factor. The relative importance of each item should not be determined using this equation because the coefficients are scaled to accommodate the units of each factor and the intercept is not at the center of the design space.

14.6 Diagnostics plots

Raw residuals and internally studentized options are also available, with externally studentized residuals being the default. The standard errors of the residuals are different unless all the runs in a design have the same leverage. Each raw residual represents a different population (one for each different standard error). As a result, it is not recommended to validate the regression assumptions using raw residuals. All of the individual normal distributions are mapped by studentizing the residuals to a single standard normal distribution. The default is externally studentized residuals based on a deletion procedure since they are more sensitive to detecting issues with the analysis. Internally Studentized residuals are also available but are less sensitive to finding such problems. As described in the diagnostics plot in **Figure 12** from design expert software.

Biodiesel Yield	=
+80.46	
−9.45	* A
−3.20	* B
+2.13	* C
−5.54	* A^2
−15.97	* B^2
−5.65	* C^2

Table 19.
Equation at the end using coded factors.

Biodiesel Yield	=
−902.92058	
+26.92040	* Temperature
+12.34881	* Methanol-Oil ratio
+19.01074	* Catalyst Weight
−0.221613	* Temperature^2
−0.283914	* Methanol-Oil ratio^2
−2.51265	* Catalyst Weight^2

Table 20.
Final equation using actual factors.

Normal Probability: If the residuals follow a normal distribution, they should follow a straight line, according to the normal probability plot. Even with typical data, expect some scatter. Only focus on distinct patterns, such as an "S-shaped" curve, which suggests that a response modification might lead to a more accurate analysis.

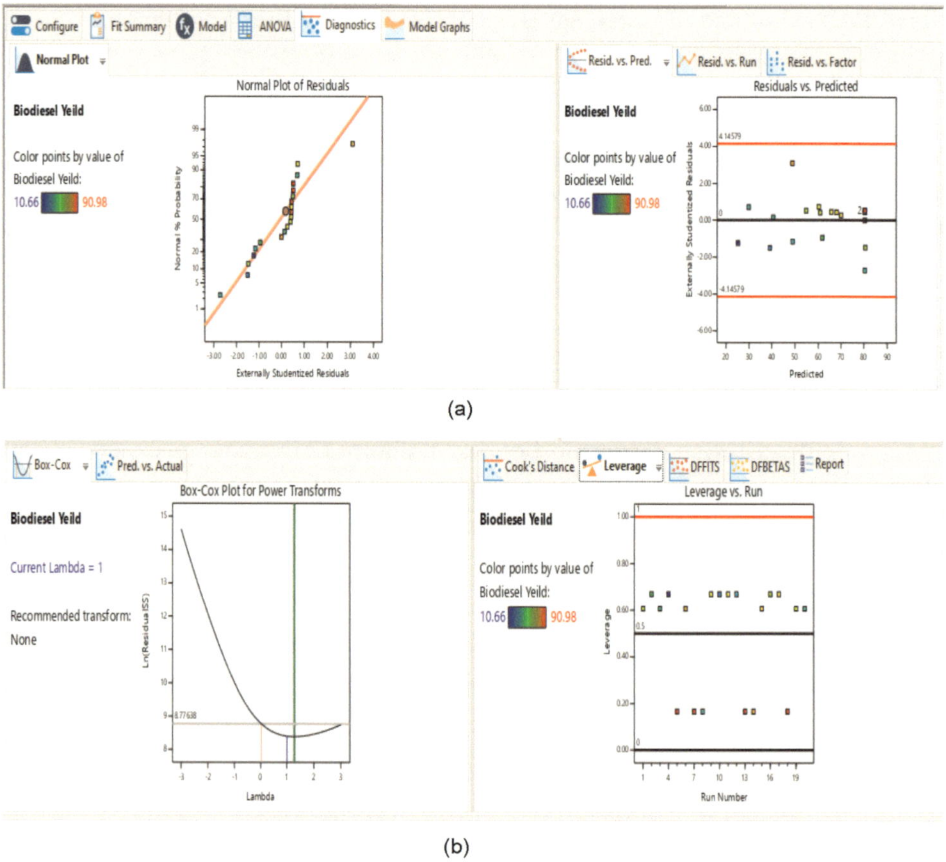

Figure 12.
A diagnostics plots.

Residuals vs. Predicted: This is a plot of the residuals versus the ascending predicted response values. The idea of constant variance is tested. The plot needs to be random scatter (residuals should have a constant range across the graph). This plot's expanding variance ("megaphone pattern") suggests that a transformation is required.

Predicted vs. Actual: An illustration showing a graph of expected and actual response values. The purpose is to detect a value, or group of values, that are not easily predicted by the model.

Leverage: A measurement of each point's impact on the model's fit. When a point's leverage is 1, the model perfectly describes the observation at that location. The model is influenced by that point. A run with more than two times the typical leverage is generally regarded as having high leverage. There aren't many runs like them in the factor space. The average leverage is calculated by dividing the number of terms among the model by the number of design runs.

14.7 Model graphs

All the model graphs which can be used to drive insights on the responses for all input data are shown in **Figures 13–18** respectively.

15. Optimization

Here, Our goal is to maximize Biodiesel Yield using the given factors in the range (lower and upper level) summarized in **Table 21** below.

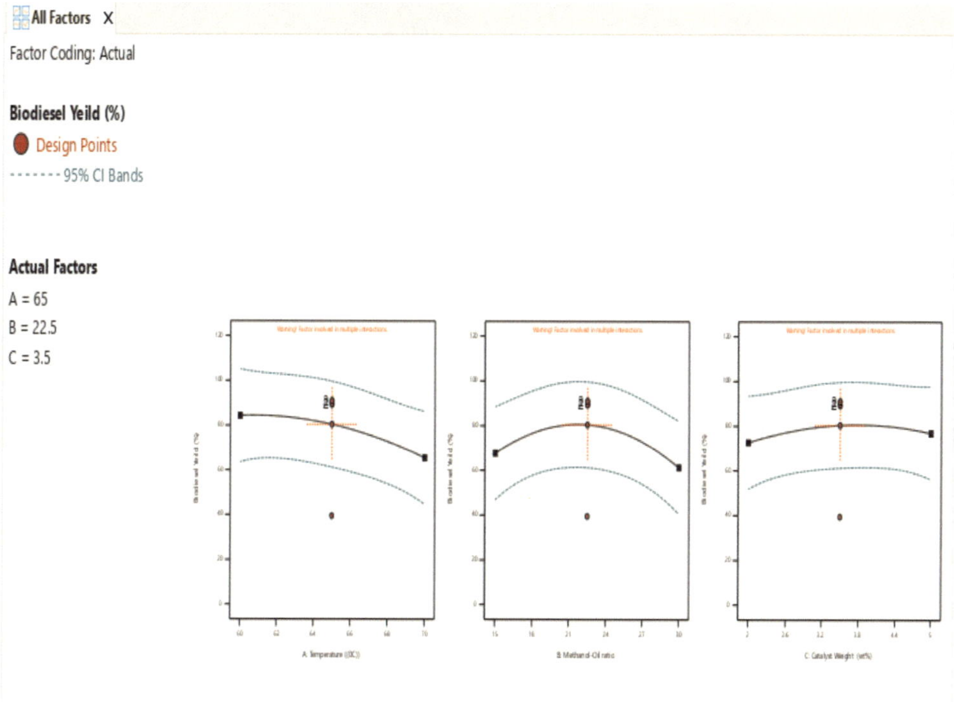

Figure 13.
All factors response.

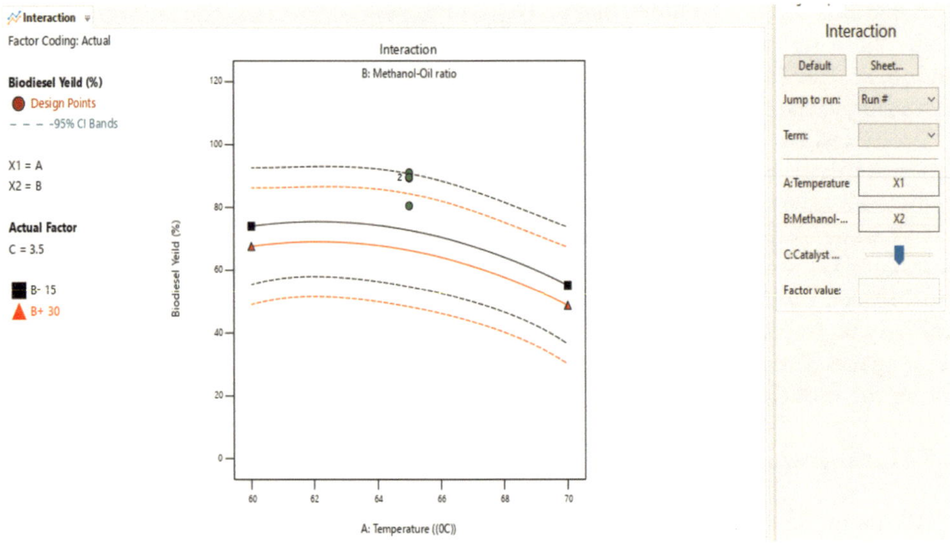

Figure 14.
Interactions.

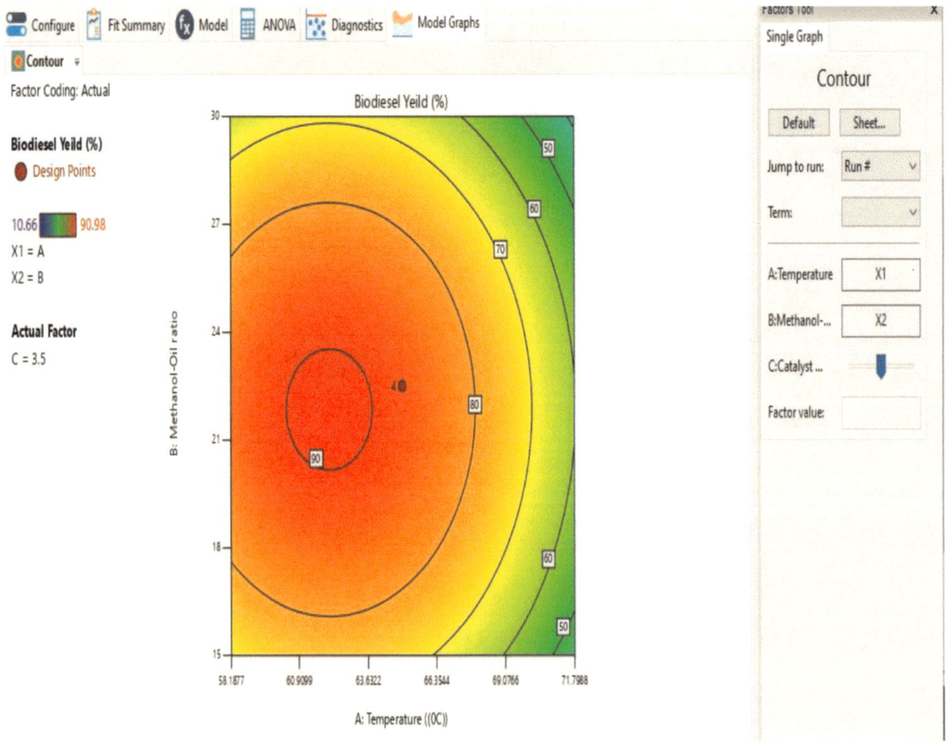

Figure 15.
Contour plot.

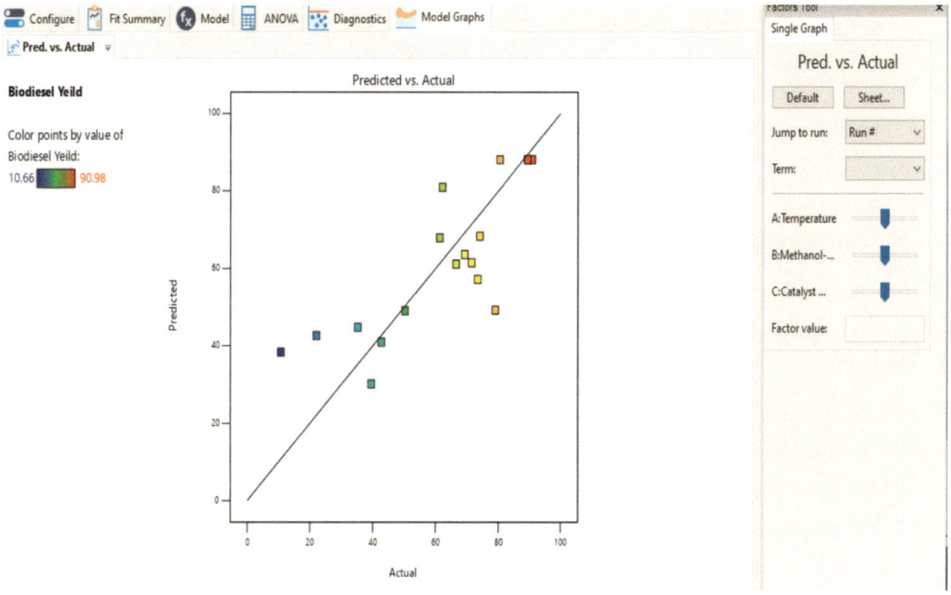

Figure 16.
Predicted vs. actual.

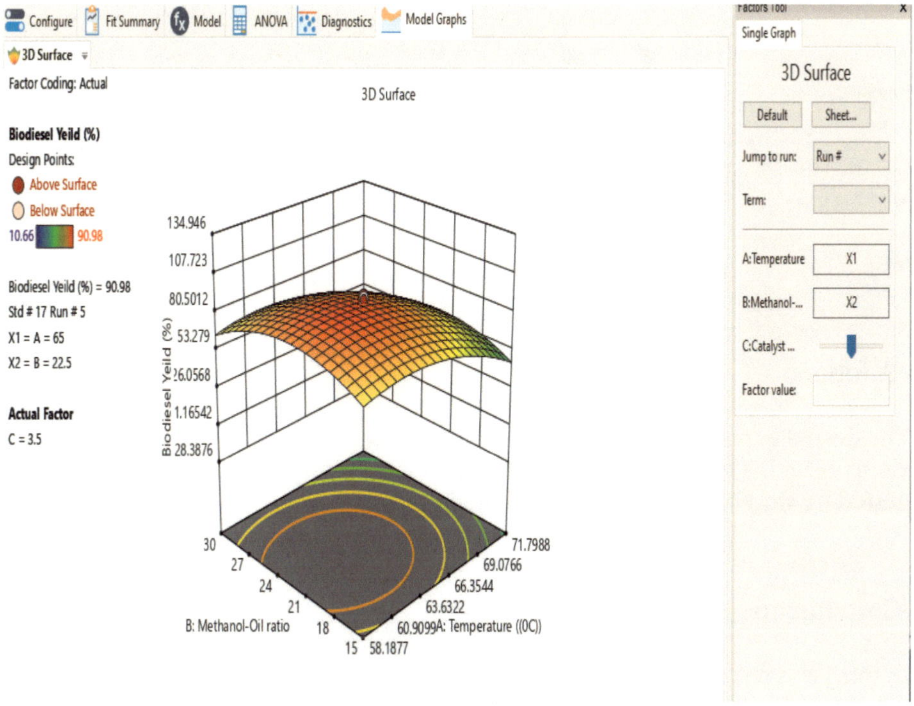

Figure 17.
3D surface plot.

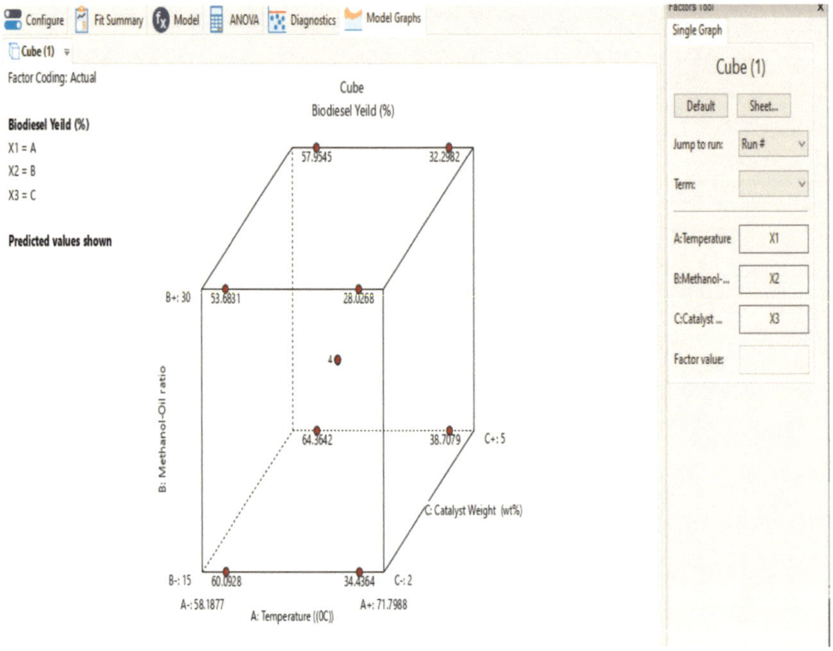

Figure 18.
Cube plot.

Name	Goal	Lower Limit	Upper Limit	Lower Weight	Upper Weight	Importance
A:Temperature	is in range	60	70	1	1	3
B:Methanol-Oil ratio	is in range	15	30	1	1	3
C:Catalyst Weight	is in range	2	5	1	1	3
Biodiesel Yield	maximize	10.66	90.98	1	1	5

Table 21.
Constraints.

15.1 Solutions

The design expert Software iterated over all the ranges of factors and found the maximum yield. There are 100 possible solutions. However, we will select the one suggested by the software and shown below in **Table 22**.

16. Conclusion

In this Chapter, we have extensively applied Central Composite design to optimize Biodiesel Synthesis Using a Catalyst and design expert 13 has been used to provide deep statistical analysis. A reduced Quadratic model with a significant p-value of 0.0325 was accepted since the Quadratic model has an insignificant p-value. The

100 Solutions found						
Number	**Temperature**	**Methanol-Oil ratio**	**Catalyst Weight**	**Biodiesel Yield**	**Desirability**	
1	61.818	22.128	3.760	91.007	1.000	Selected
2	62.522	21.568	3.704	90.987	1.000	
3	62.223	21.803	3.668	91.064	1.000	
4	61.717	21.882	3.708	91.025	1.000	
5	62.126	21.973	3.759	91.052	1.000	
6	62.021	21.746	3.772	91.045	1.000	
7	62.295	21.826	3.807	91.013	1.000	
8	62.025	22.087	3.706	91.052	1.000	
9	62.183	21.870	3.759	91.055	1.000	
10	62.188	22.001	3.632	91.049	1.000	

Table 22.
Optimization solutions.

model is significant, as indicated by the model's F-value of 3.57. An F-value this large might be caused by noise only in 3.25% of cases. The number of the experimental run was reduced to 18 runs compared to the 20 runs used by the original experimenters and we have also obtained a higher yield of 91% compared to the 89% obtained in the original study.

Acknowledgements

I acknowledge my co-author, Dr. C.N Njoku for his help and support to garner this information and for inspiring the success of this work. I also want to return big regard to my mother for always providing her special support in the little ways she could.

Author details

Chigoziri N. Njoku[1] and Samuel K. Otisi[2]*

1 Africa Centre of Excellence in Future Energies and Electrochemical Systems (ACE-FUELS), Federal University of Technology, Owerri, Nigeria

2 Department of Chemical Engineering, Federal University of Technology, Owerri, Nigeria

*Address all correspondence to: samuelotisikalu@gmail.com

IntechOpen

© 2023 The Author(s). Licensee IntechOpen. This chapter is distributed under the terms of the Creative Commons Attribution License (http://creativecommons.org/licenses/by/3.0), which permits unrestricted use, distribution, and reproduction in any medium, provided the original work is properly cited.

References

[1] Bhattacharya S. Central composite Design for Response Surface Methodology and its Application in pharmacy. In: Response Surface Methodology in Engineering Science. London, UK: IntechOpen; 2021. DOI: 10.5772/INTECHOPEN.95835

[2] Wikipedia contributors. (2020). Central composite design. In Wikipedia, The Free Encyclopedia. Available from: https://en.wikipedia.org/w/index.php?title=Central_composite_design&oldid=954106283 [Accessed: May 14, 2022]

[3] Skartland LK, Mjos SA, Grung B. Experimental designs for modeling retention patterns and separation efficiency in the analysis of fatty acid methyl esters by gas chromatography-mass spectrometry. Journal of Chromatography A. 2011;**1218**: 6823-6831

[4] Tshizanga N, Aransiola EF, Oyekola O. Optimization of biodiesel production from waste vegetable oil and eggshell ash. South African Journal of Chemical Engineering. 2017;**23**:145-156. DOI: 10.1016/j.sajce.2017.05.003

[5] Manohar M, Joseph J, Selvaraj T, Sivakumar D. Application of box Behnken design to optimize the parameters for turning Inconel 718 using coated carbide tools. International Journal of Scientific and Engineering Research. 2013;**4**(620):642

[6] Breyfogle FW. Chapter 17. In: Statistical Methods for Testing, Development, and Manufacturing. John Wiley & Sons Ltd, New York. 252 p; 1992

[7] Singh B, Kumar R, Ahuja N. Optimizing drug delivery systems using systematic" design of experiments." part I: Fundamental aspects. Critical Reviews in Therapeutic Drug Carrier Systems. 2005;**22**(1):27-105

[8] Cavazzuti M. Design of experiments. In: Optimization Methods. Berlin, Heidelberg: Springer; 2013. pp. 13-42

[9] Hassanein HM, Abd-Rabou AS, Sakr SM. Design optimization of transverse flux linear motor for weight reduction and performance improvement using response surface methodology and genetic algorithms. IEEE Transactions on Energy Conservation. 2010;**25**(3): 598-605

[10] Anderson MJ, Whitcomb PJ. RSM Simplified. New York: Productivity, Inc.; 2016

[11] DeGryze L, Vandebroek. Using the correct intervals for prediction: A tutorial on tolerance intervals of ordinary least-squares regression. Chemometrics and Intelligent Laboratory Systems. 2007;**87**(2):147-154

[12] Zahran A, Anderson-Cook CM, Myers RH. Fraction of design space to assess prediction capability of response surface designs. Journal of Quality Technology. 2003;**35**(4):377-386

Chapter 3

Applications of Response Surface Methodology (RSM) in Product Design, Development, and Process Optimization

Sheriff Lamidi, Nurudeen Olaleye, Yakub Bankole, Aishat Obalola, Emmanuella Aribike and Idris Adigun

Abstract

In this review chapter, the authors presented a systematic exposition to the concept of Response Surface Methodology (RSM) for applications by Scientists, Engineers, Technologists and Industries. (RSM) is an empirical model which employs the use of mathematical and statistical techniques in relating input variables otherwise known as factors to the response. RSM became very useful due to the fact that other methods available such as the theoretical model could be very cumbersome to use, time-consuming, inefficient, error prone and unreliable. In order to draw meaningful conclusions and findings, an experiment is required. In an effort to obtain an objective conclusion (between the factors and the response), an experimenter needs to plan and design the experiments, and analyze the results. An approximation of the response in relation to the variables is otherwise known as RSM. This chapter reviews RSM concept for easy understanding and adoption by researchers. In section 2.0, the various terminologies used in RSM were defined. In section 3.0, RSM design types were highlighted and RSM research phases exposed in section 4.0. Section 8.0 gave some scenario applications of RSM in various fields and section 9.0 defined the RSM research cycle process. General applications and conclusions stated.

Keywords: response surface methodology, RSM design, optimization, RSM applications, product design

1. Introduction

Experimentation, Data collection, Data processing, and Analysis of data are very basic and essential to Scientists, Engineers, Technologists, and Manufacturing Industries to design, develop, improve and validate their products, processes, and operations. Response surface methodology (RSM) which is available in MINITAB and other proprietary software is a collection of both statistical and mathematical techniques useful for developing, improving, and optimizing processes [1]. RSM is known to play a pivotal role in new product design and development as well as in improving existing ones. With response surface methodology we can determine the optimum factor needed to produce the best result. RSM is a critical and very robust tool for data manipulation and analysis of research data to obtain a quality result or an improvement [1]. RSM could be applied by an industry that desires to manufacture a component (from Al-Si Alloy material) with minimum surface roughness by combining three controllable variables (cutting speed, feed rate, and, depth of cut). Because of this, the Design of Experiments (DOE) could be used to carry out the study of the effect of the three machining variables (cutting speed, feed rate, and depth of cut) on the surface roughness (Ra) of Al-Si alloy [2]. With the use of response surface methodology (RSM), a mathematical prediction model of the surface roughness would be developed in terms of cutting speed, feed rate, and depth of cut. The effects of the three process parameters on both Ra can then be investigated by using the response surface methodology (RSM). The above approach can be adopted by any industry, scientist, or researcher in getting better results (response) from several variables otherwise known as factors. RSM helps to reduce the noise in an experiment, thereby ensuring optimization. Many researchers have conducted researches on the application of RSM or other DOE concept in which the results of their findings have been used to develop a predictive model in several fields such as; tool life modeling, surface roughness prediction, for monitoring and functionality or health condition of electronic devices also for the surface roughness of Inconel using full factorial design of experiment among other areas of applications [2–4]. The RSM looks into an adequate approximation relationship between input and output variables and determines the best operating circumstances for a system under study or a portion of the factor field that complies with the operating requirements or conditions [3, 5, 6].

Response surface methodology can be better referred to as a collection of statistical and mathematical techniques employed for product design and improvement, process development and improvement as well as process optimization. It has major applications in the design, development, and, formulation of new products as well as in improving existing product design. RSM is a robust tool for the design of experiments, analysis of experimental data, and process optimization. In RSM, the response is determined by the variables and the aim is to optimize the response [1, 7, 8]. There are two primary experimental designs used in response surface methodology: Box-Behnken designs (BBD) and central composite designs (CCD) [8, 9]. Recently, optimization studies have also used central composite rotatable design (CCRD) and face central composite design (FCCD) [8, 10–14].

Wong [15] employed RSM concept to carry out reliability analysis of soil slopes. Tandjiria et al. [16] used response surface method for reliability analysis of laterally loaded piles. Sivakumar Babu and Amit Srivastava [17] presented a study on the analysis of allowable bearing pressures on shallow foundation using response surface method and showed that a comparative study of the results of the analysis from conventional solution and numerical analysis in terms of reliability indices enables rational choice of allowable loads. For better understanding of the RSM concept in our daily life experiences as described in **Figure 1**. Take for example we have two variables (humidity and temperature) and we want to see the effects of these variables on human comfort. We can name these independent variables temperature and humidity, X1 and X2 and the response which is human comfort can be named Y. Response Surface Methodology is useful in this case for the **modeling** and optimization of the situation above in which the

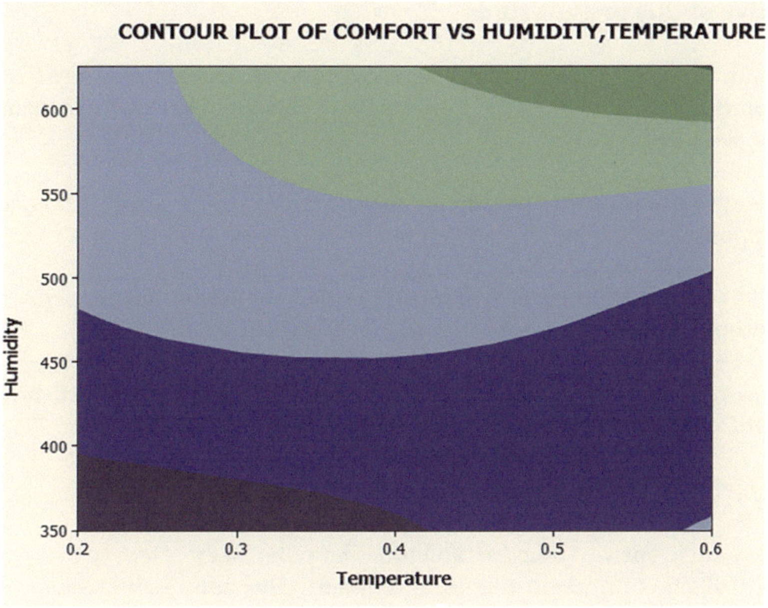

Figure 1.
Response surface for humidity and Temperature on human comfort.

response of interest (human comfort) is influenced by the **variables (humidity and temperature).** In this model example, our objective is to **optimize** this response The **visual representation** of the above is otherwise known as Response Surface Methodology (RSM) or response surface modeling. To find the levels of temperature (X1) and pressure (X2) for maximum human comfort (y) in the above process.

$$y = [(x_1,\ x_2) + \epsilon] \tag{1}$$

ϵ is referred to as the error term inherent in the system

1.1 The concept of RSM

The concept of Response Surface Methodology can be used to establish an approximate explicit functional relationship between input random variables and output response through regression analysis and probabilistic analysis can be performed [15]. RSM involves a combination of metamodeling (i.e., regression) and sequential procedures (iterative optimization). Response Surface Methodology (RSM) is a collection of mathematical and statistical techniques useful for the modeling and analysis of problems. By careful design of experiments, the objective is to optimize a response (output variable) that is influenced by several independent variables (input variables). A collection of mathematical and statistical methods called Response Surface Methodology (RSM) can be used to simulate and analyze issues. The goal of meticulous experiment design is to maximize a response (output variable) that is affected by a number of independent variables (input variables). The motivation behind this work is the applicability of the concept of RSM to many areas of scientific research, engineering and manufacturing industries.

1.2 Objective of this present study

The applications of RSM is for product and process development are discussed through some general and scenario applications. The chapter review presented, is shown that with RSM we can;

i. identify the sensitive parameter that provides the greatest influence on the response.

ii. easily take decision that will impact positively the product design and process optimization.

iii. ensure reliability, acceptability and profitability of the product developed and/or optimized condition.

2. Some useful terminologies

- **Factors** are input variables/parameters that potentially affect the response. It can be controllable or uncontrollable, and quantitative and qualitative.
- **Response** is a dependent variable. It is the desired results obtained from combining the interaction of independent variables.
- **Experiment** is a series of tests, called runs, in which changes are made in the input variables to identify the reasons for changes in the output response.
- **Experimenter is** a person experimenting for research purposes.
- **Treatment** is a combination of one or more factors.
- **Levels** are the values a factor can take on
- **Effect** simply means how much a main factor or interaction between two or more factors influences the response.
- Design Points: simply means the assigned values of the individual factors for which the experiment was performed.
    - One design point = one treatment
    - Points are typically coded to more practical values.
    - example. 1 factor with 2 levels – levels coded as (−1) for low level and (+1) for high level
- **Design Space** is the range of values over which factors are to be varied or adjusted [18, 19].
- **Response Surface** is the unknown or experimental purpose. It is the mean response at any given level of the factors in the design space.

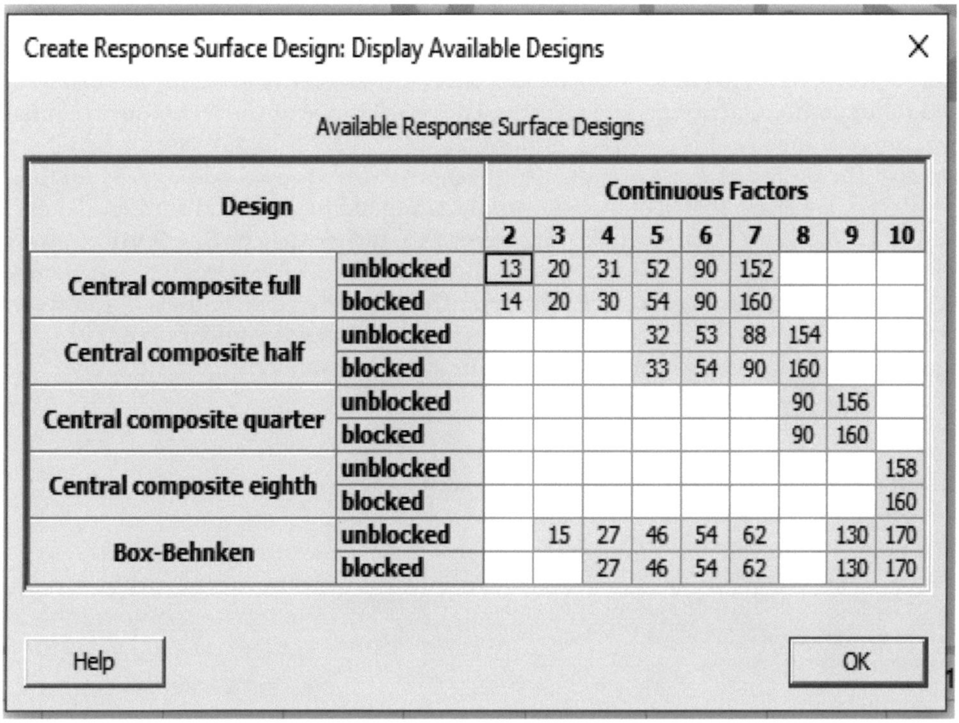

Design		Continuous Factors								
		2	3	4	5	6	7	8	9	10
Central composite full	unblocked	13	20	31	52	90	152			
	blocked	14	20	30	54	90	160			
Central composite half	unblocked				32	53	88	154		
	blocked				33	54	90	160		
Central composite quarter	unblocked							90	156	
	blocked							90	160	
Central composite eighth	unblocked									158
	blocked									160
Box-Behnken	unblocked		15	27	46	54	62		130	170
	blocked			27	46	54	62		130	170

Figure 2.
Available Designs in RSM source MINITAB 20.

RSM Design Types; The summary of the various types of design available in response surface methodology is presented in **Figure 2** according to [18].

(ii) Central Composite Design (CCD) (2 to 10 continuous factors)

(ii) Box-Behnken Design (3,4,5,6,7,9 or 10 continuous factors)

To do a visual analysis of the response surface design, the designer can use the following visualization tool to visualize the response in RSM.

- Residual plots
- Effect plots

3. RSM research phases

RSM involves four broad phases as highlighted below.

i. Use a simulation model e.g., Minitab to fit a linear regression model to the data points in the workspace and, find a better solution from the linear regression model.

ii. Repeat the above process until the slope of the linear response surface obtained from the linear regression model is approximately zero.

iii. Fit a nonlinear quadratic regression.

iv. Lastly find the optimum of this equation.

4. Getting access to response surface methodology (RSM)?

MINITAB, STATISTICA, DESIGN EXPERT, etc. are software tools that can be used for experimental design and analyze data. RSM is one of the techniques that have been programmed in this software. Among all, MINITAB is highly rated when it comes to the design of experiments using response surface methodology. Minitab is a proprietary software tool, a computer program applied in statistical studies, developed in 1972. Its interface is similar to Microsoft Excel or Calc of OpenOffice, used in universities and companies, it has specific functions focused on process management and analysis of the Six Sigma suite. Minitab offers Quality Control tools, Experiment Planning (DOE) e.g., RSM, Reliability Analysis, and General Statistics [18, 20]. **Figure 3** shows the navigation process in Minitab 18 to access response surface methodology interface (**Table 1**).

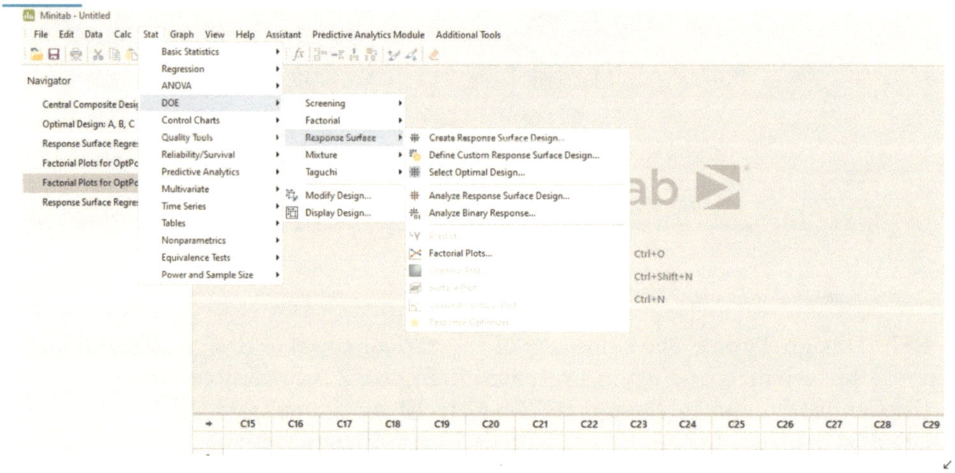

Figure 3.
Diagram showing the navigation of RSM with MINITAB software.

S/N	Factors and types	Responses
1.	mean interactive time (uncontrollable, quantitative)	Mean daily production rate
2.	Mean service time (controllable, uncontrollable, quantitative)	Meantime in the system for patient
3.	Number of servers (controllable, quantitative)	Mean inventory level
4.	Reorder point (controllable, quantitative)	Mean surface roughness
5.	Queuing discipline, qualitative)	Number of students who wait till the end of the lecture
6.	Mean winter demand time (uncontrollable, quantitative)	Mean material removal rate

Table 1.
Some examples of Factors and Response in RSM.

5. Advantages of RSM

The application of response surface methodology in research and industry comes with the following advantages

i. Seamless statistical analysis

ii. Optimization of manufacturing system, process, or product.

iii. Experimental layout and design.

iv. Prediction

v. Interaction of variables is easily presented with clear curves and other visual aids.

vi. Good visualization of responses or results with the use of surface plots, graphs, etc.

vii. Associated empirical mathematical models.

6. Some scenarios of RSM Applications

6.1 A scenario of RSM in a manufacturing process

In a quest to manufacture a component during CNC turning operation. Proper selection of process parameters or variables (cutting speed, feed rate, and depth of cut) for optimal surface quality (Response) must be achieved. This requires a more methodical approach by using experimental methods and mathematical and statistical models. The design of experiments will play a pivotal role in this regard. This will require considerable knowledge and experience of the designer to design experiments and analyze data. Note that the traditional design-of-experiment (DOE) technique requires a large number of samples to be produced. To increase machining process efficiency, strategies for optimizing machining parameters using experimental methodologies as well as mathematical and statistical models have developed significantly over time. A full factorial approach may be required to look into all potential combinations to build an approximation model that can describe interactions between design variables in this CNC turning operation. An experimental approach known as a factorial experiment involves varying design variables simultaneously rather than one at a time. It is necessary to define the lower and upper bounds for each of the n design variables in the optimization problem. Then, at various levels, the permitted range is discounted. If just the lower and upper bounds (two levels) of each variable are defined. The experimental design is referred to as 2n full factorial if each variable is defined at just the upper and lower boundaries (two levels). Second-order models can be fitted using factorial designs. When a first-order model exhibits a lack of fit as a result of the interaction between variables and surface curvature, a second-order model can considerably enhance the optimization procedure. The goal of a meticulous experiment design is to optimize the response. (Surface quality of the machined part) which is influenced by several independent input variables (cutting speed, feed rate, and depth of cut).

6.2 A scenario of RSM in the energy industry

Due to the limited availability of high-grade coal for energy production, low-grade coal can be employed. High ash levels and high moisture content are characteristics of low-grade coal. With the use of the response surface methodology, the operational parameters were optimized to generate clean coal as effectively as possible. The impact of three independent variables, including hydrofluoric acid (HF) concentration (10–20 percent by volume), temperature (60–100°C), and time (90–180 min), for ash reduction from the low-grade coal, was explored to attain this coal optimization target. By utilizing the central composite design (CCD) method, a quadratic model was presented to correlate the independent variables for maximal ash reduction at the ideal process condition. In comparison to time and temperature, the study finds that HF concentration was the most efficient parameter for ash reduction [16].

6.3 A scenario of RSM in extraction optimization

In order to maximize the extraction process of oil from leaves, fruits etc., it is important to optimize the extraction parameters so as to get the best yield. RSM concept has been used more often in recent years to optimize various oil extractions from plant sources [17, 21].

6.4 A scenario of RSM in drinking water treatment process

Both trihalomethanes (THMs) and Natural Organic Matter (NOM) has been characterized with cancer risk in drinking water According to [22]. The concept of RSM was used for the development of water treatment technologies and optimization of process variables in order to reduce THMs and NOM level of concentration in drinking water. A model was developed to control the process. The developed models can be effectively used to remove both THMs and NOM from drinking water.

6.5 A scenario of RSM in construction industry

The construction industry is a very germane industry in the technological advancement of any nation. The level of research-based construction has been improved lately. A study on the analysis of allowable bearing pressures on shallow foundation using response surface method was conducted and showed that a comparative study of the results of the analysis from conventional solution and numerical analysis in terms of reliability indices enables rational choice of allowable loads [15].

6.6 A scenario of RSM in product development

The effect of oven parameters such as air velocity, time, temperature etc. on formulations (sugar, water, fats, flavors, etc.,) of the quality of baked food product can the analyzed with the application of response surface methodology [23]. RSM model is a powerful tool to optimize the product quality (volume of baked product, crust and crumb color, bake loss among others). The data collected through RSM can further be used to obtain the variability of the response(s) with tested parameters [23]. In this scenario, the results of the optimization obtained is otherwise referred to as quality product.

RSM cycle processes is shown in **Figure 4**.

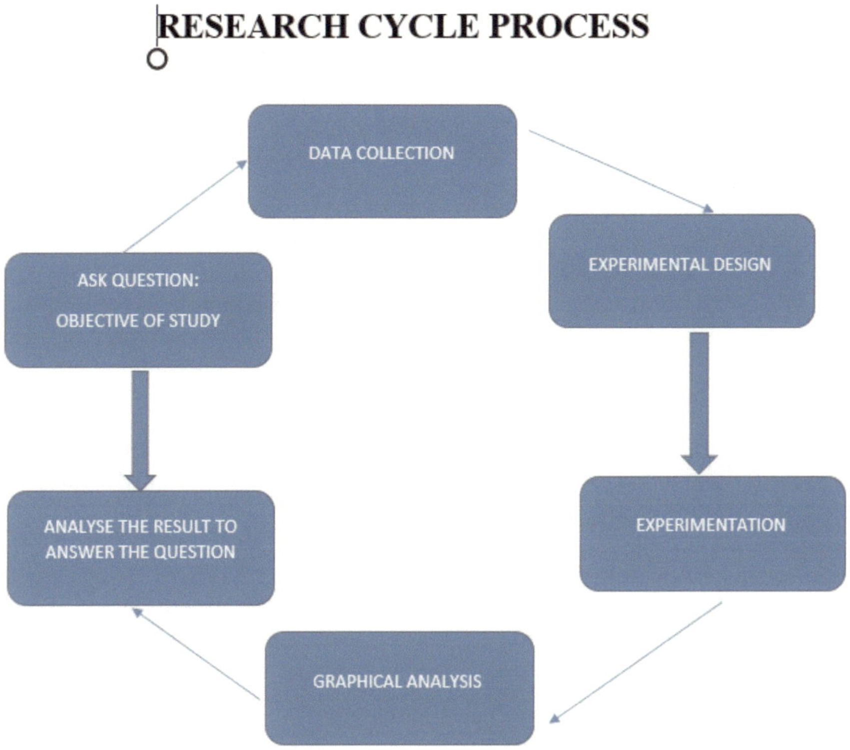

Figure 4.
RSM Research Cycle Process.

1. **Experimental Plan/Data collection.** This is the initial stage. The planning session precedes any other session involved in RSM modeling. In this session, all decisions involved in the project or experiment are clearly stated and defined. Some of the decisions under this heading include the research objective, methodology of the research, and variables that could influence the results are highlighted. This process takes care of all necessary information regarding the experimental strategy [20, 24–26]. To clarify the objective of the experiment, the objective must determine;

    a. What data is to be collected?

    b. How to measure it?

    c. How does the data relate to processing performances and experimental objectives?

2. **Experimental Design.** Experimental design can either be a conventional method or a statistical method. The conventional method has the following features;

    a. It is time-consuming.

    b. Can handle one factor over time (OVAT) or one factor at a time (OFAT).

    c. Interaction between two or more variables cannot be interpreted.

Features of statistical method experimental design

i. It is otherwise known as Design of Experiment (DOE)

ii. Apply the factorial concept

iii. It makes use of modeling to predict the behaviors of process variables e.g., RSM

iv. The process variables could be explained through interaction plots and graphs

v. Saves time and improves efficiency.

An experiment is designed based on the decisions during the designing or data collection stage. The experimental design clearly states the number of experiments and how the experiment will be carried out [20, 26, 27].

3. **Conducting Experiment.** The next step after the experimental design is to experiment with the exact research parameters and, in the order, defined by the layout for easy statistical validity. The person experimenting is called an experimenter.

4. **Analyzing the Results.** The primary focus in this analysis stage is to obtain useful information from the experiment conducted and ascertain the level of quality or improvement recorded. In this stage, the results obtained from the experiment conducted are analyzed. The analysis of the results is targeted toward specific conclusions. Since we have several samples tested in each experimental run, different analysis techniques can be selected.

5. **Graphical Analysis.** One of the powerful features of RM is the ability to present results or responses using visual aids for easy interpretation or understanding.

6. **Confirmation of results/Ask questions relating to the research objective.** A test can be carried out to ascertain if the actual performance of the product in-service condition matches the improvement stated in the results. The test here helps to determine the research gap (**Figure 5**).

7. General applications of RSM

i. RSM found its major application in the industry to model and design a product. Also, to optimize the manufacturing process.

ii. RSM is capable of Data analysis, prediction, product design, and optimization [28, 29].

iii. RSM can predict the relationship or interaction that exists between the values of some measurable response variable(s) and those of a set of experimental factors presumed to affect the response(s).

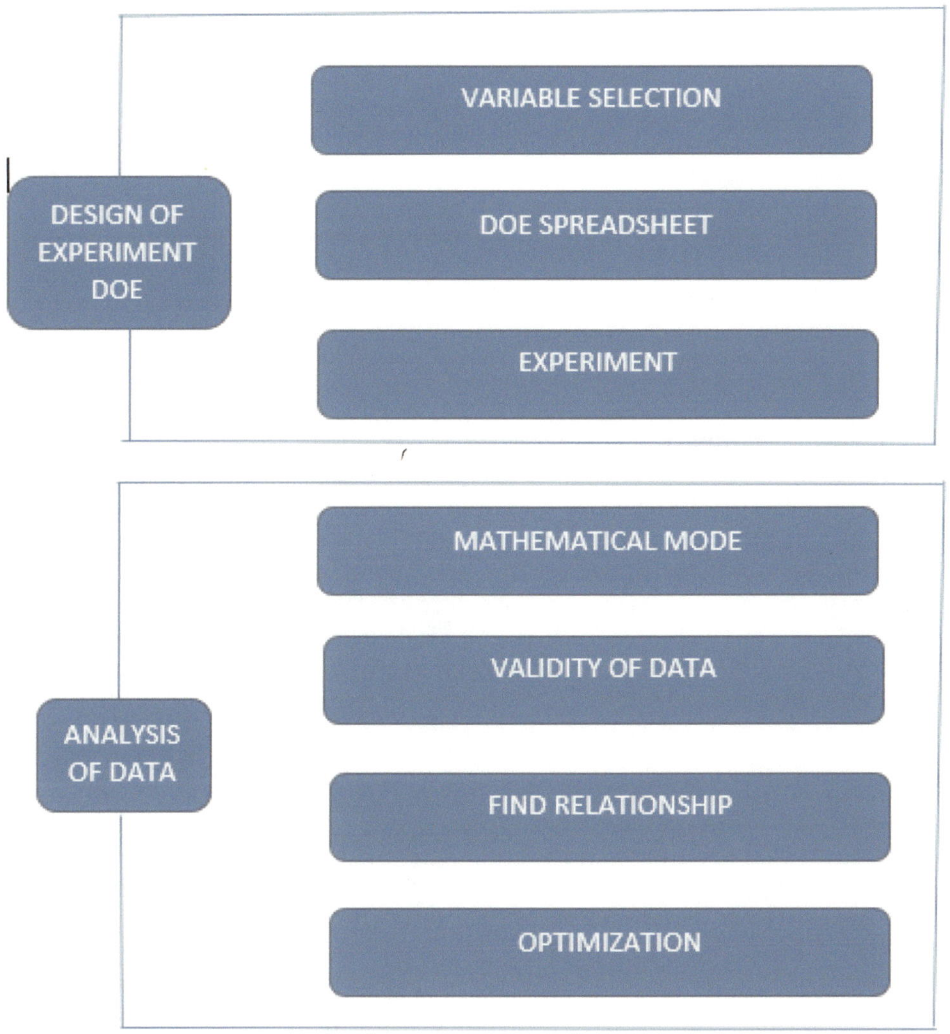

Figure 5.
RSM Flow chart.

iv. RSM is capable to predict the response value at various process conditions.

v. Application of RSM can be used to screen independent variables in order to determine most significant main effect of factors among several potential ones.

vi. With RSM a non-identified interaction effect could be determined.

vii. With RSM application for optimization, one can easily identify the best factor (s), process interaction effect, that produce the response that brings the optimized condition. This is the real deal in parameter optimization.

viii. RSM enhances product quality, product life span and increase productivity.

ix. Application of response surface methodology exposes the best type of design: CCD or BBD needed to achieve optimization.

x. RSM can predict the relationship or interaction that exists between the values of some measurable response variables and those of a set of experimental factors presumed to affect the response.

xi. RSM is capable to predict the response value at various process conditions.

xii. Power visual aids such as plots graphs in 2D or 3D for easy understanding of the results.

8. Visualizing results in RSM

RSM uses a variety of surface visualization techniques according to **Figures 6–11** to visually assess how factors affect the response. When a regression model is fitted as a result of interactions between two or more predictors, visualization better communicates the experimental results or responses. Effects plots, contour plots, residual plots, surface plots, etc. are a few examples of graphical visualization tools also known as response surface plots. These plots aid in determining the process conditions and desired response Values [18, 19, 27, 29–35].

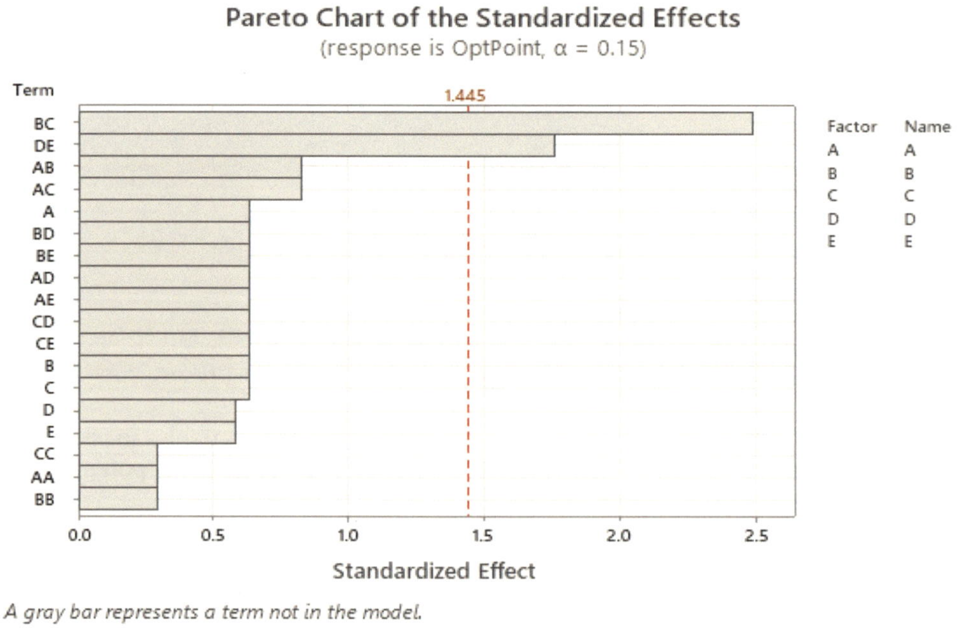

Figure 6.
Figure showing standardized effect.

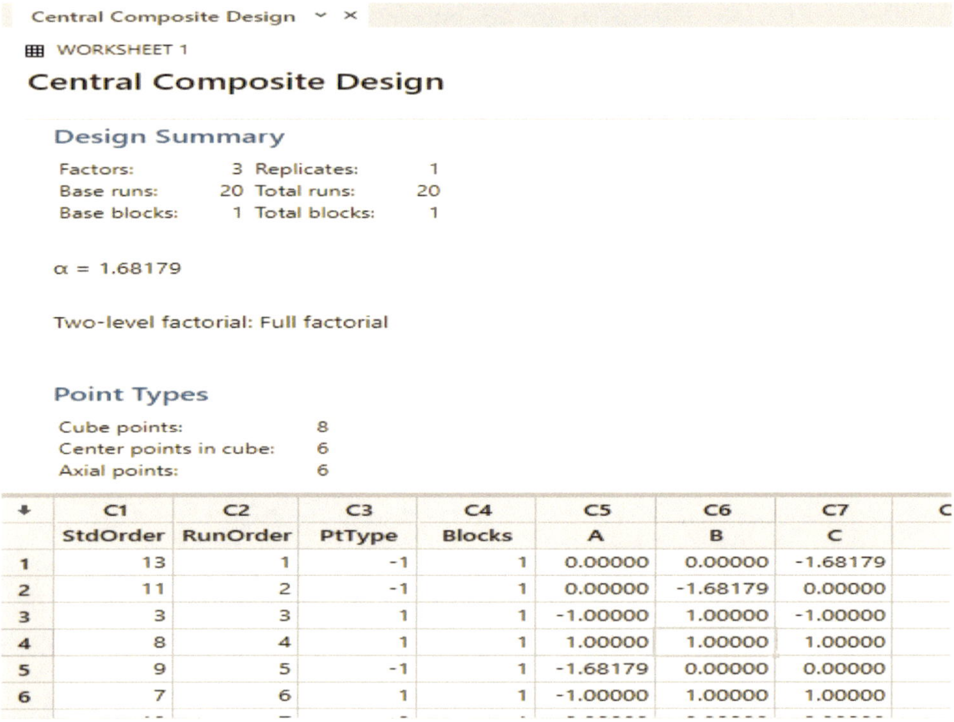

Central Composite Design

WORKSHEET 1

Central Composite Design

Design Summary

Factors:	3	Replicates:	1
Base runs:	20	Total runs:	20
Base blocks:	1	Total blocks:	1

α = 1.68179

Two-level factorial: Full factorial

Point Types

Cube points:	8
Center points in cube:	6
Axial points:	6

↓	C1	C2	C3	C4	C5	C6	C7
	StdOrder	RunOrder	PtType	Blocks	A	B	C
1	13	1	-1	1	0.00000	0.00000	-1.68179
2	11	2	-1	1	0.00000	-1.68179	0.00000
3	3	3	1	1	-1.00000	1.00000	-1.00000
4	8	4	1	1	1.00000	1.00000	1.00000
5	9	5	-1	1	-1.68179	0.00000	0.00000
6	7	6	1	1	-1.00000	1.00000	1.00000

Figure 7.
Central composite design.

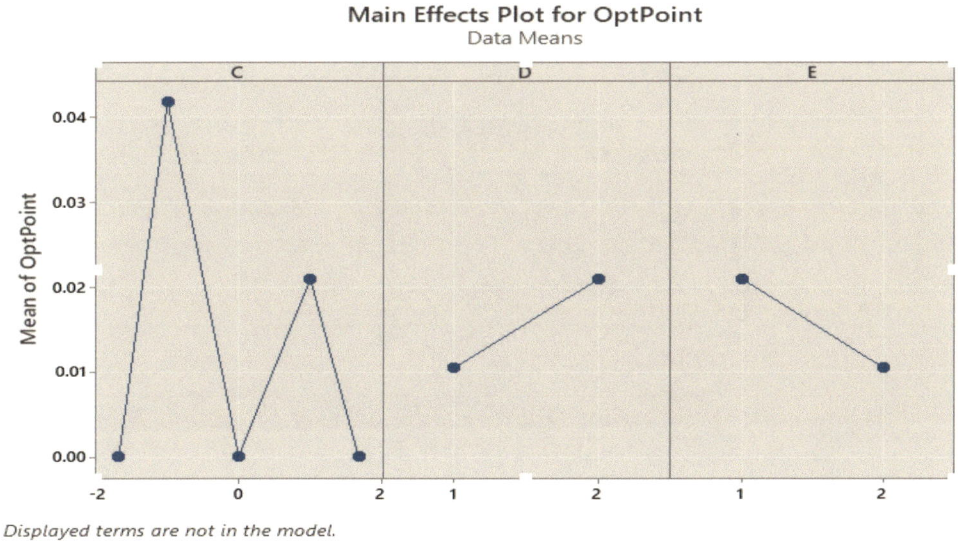

Figure 8.
Main effect plot in RSM.

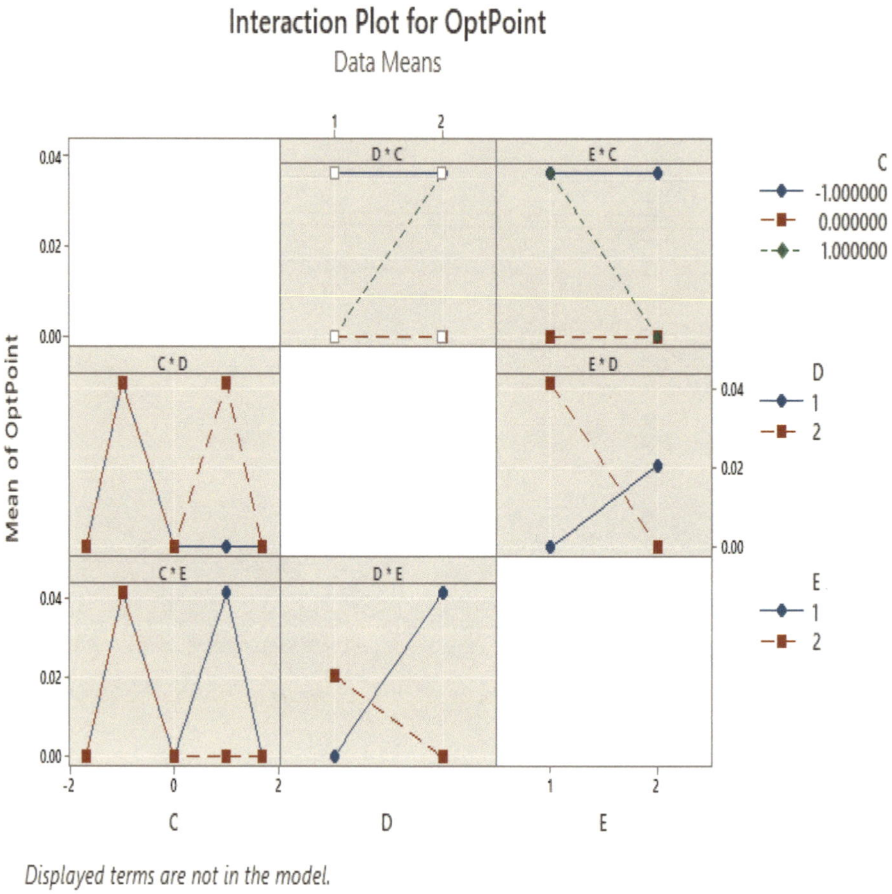

Figure 9.
Interaction plot explored from RSM software.

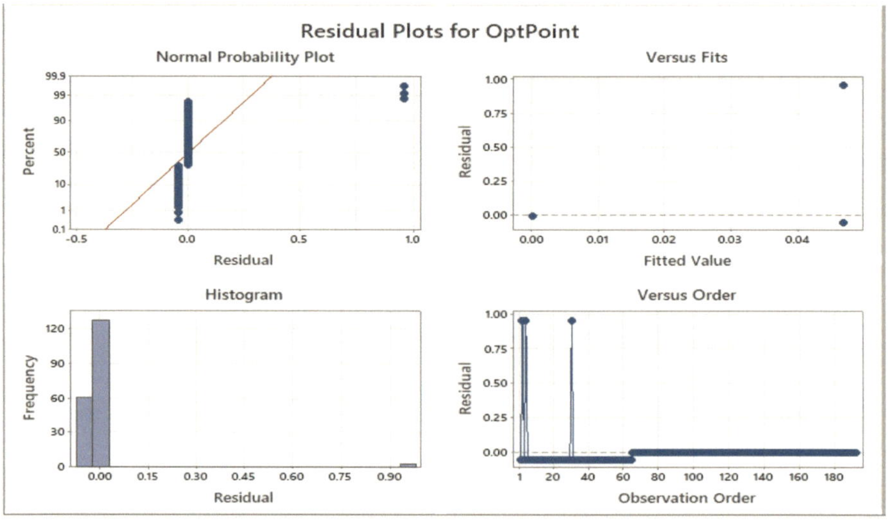

Figure 10.
Residual plot for optimized point.

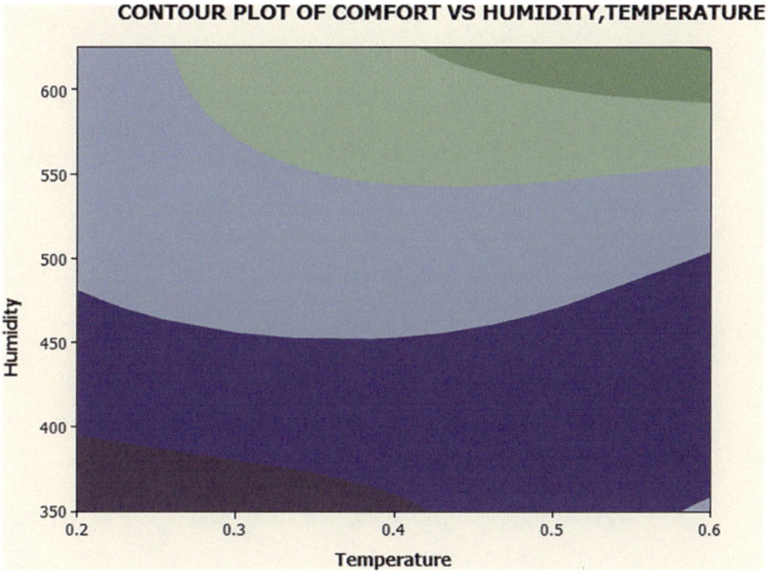

Figure 11.
Some response surface plots for visualization of RSM results.

9. Conclusions

In this chapter, the authors provided a detailed approach for the understanding and implementing Response Surface Methodology (RSM) for the various professionals or researchers who may be involved in the application of Response Surface Methodology. In an attempt to design a product or to optimize an existing process there are several methods that can be adopted. RSM has many advantages when compared to classical methods. It requires fewer runs of experiments to understand the effects of all the factors and the optimum combination of all factor input. RSM requires less time, removes trial by error and ensure high quality results. Having presented in this chapter the huge applications of RSM in various fields of research, it can be concluded that RSM is a great research tool for product design, development and process optimization. The chapter coverage is detailed enough to give the basic insight of RSM even to a novice hearing about RSM for the very first time. However, the chapter does not cover all the information required for mastery of the RSM concept.

Acknowledgements

The authors sincerely appreciates IntechOpen with the opportunity provided to publish this chapter review in the main book; Response Surface Methodology-Research Advances and Applications.

We also give kudos to the critical peer review process.

Notes/thanks/other declarations

I hereby testify to the good work by Intechopen by way of contributing to research across all the fields from different part of the world. Indeed, there is no better way for the advancement of research than you are doing. Please, keep it up.

Author details

Sheriff Lamidi*, Nurudeen Olaleye, Yakub Bankole, Aishat Obalola, Emmanuella Aribike and Idris Adigun
Lagos State University of Science and Technology (Formerly Lagos State Polytechnic), Lagos, Nigeria

*Address all correspondence to: lamidi.s@mylaspotech.edu.ng

IntechOpen

© 2022 The Author(s). Licensee IntechOpen. This chapter is distributed under the terms of the Creative Commons Attribution License (http://creativecommons.org/licenses/by/3.0), which permits unrestricted use, distribution, and reproduction in any medium, provided the original work is properly cited.

References

[1] Raymond H, Montgomery DC, Anderson-Cook CM. Process and Product Optimization Using Designed Experiments. Third ed. New Jersey: John Wiley & Sons, Inc; 2009

[2] Kasali Aderinmoye A, Sheriff Babatunde L. Effect of dry cutting system on surface finish in turning operation of Al-Si alloy. Journal of Multidisciplinary Engineering Science and Technology. 2021;**8**:10

[3] Hascalik A, Caydas U. Optimization of turning parameters for surface roughness and tool life based on the Taguchi method. International Journal Advance Manufacturing Technology. 2008;**38**:896-903

[4] Mead R, Pike DJ. A review of response surface methodology from biometric viewpoint. A literature survey. Biometrics. 1975;**31**:803-851

[5] Hill WJ, Hunter WG. A review of response surface methodology: A literature survey. Technometrics. 1966; **4**:571-590

[6] Peasura P. Application of response surface methodology for modeling of postweld heat treatment process in a pressure vessel steel ASTM A516 Grade 70. The Scientific World Journal. 2015

[7] Farooq Z, Rehman S, Abid M. Application of response surface methodology to optimize composite flour for the production and enhanced storability of leavened flat bread (Naan). Journal of Food Processing and Preservation. 2013;**37**:939-945

[8] Pishgar-Komleh SH, Keyhani A, Msm R, Jafari A. Application of response surface methodology for optimization of Picker-Husker harvesting losses in corn seed. Iranica Journal of Energy and Environment. 2012;**3**(2):134-142

[9] Montgomery DC. Design and Analysis of Experiment. Fifth. ed. New York, USA: Wiley, Inc.; 1997

[10] Aydar AY. Utilization of response surface methodology in optimization of extraction of plant materials. Open Access. 2018

[11] Koç B, Kaymak-Ertekin F. Response surface methodology and food processing applications. Gıda. 2009;**7**:1-8

[12] Prakash Maran J, Mekala V, Manikandan S. Modeling and optimization of ultrasound-assisted extraction of polysaccharide from Cucurbita moschata. Carbohydrate Polymers. 2013;**92**:2018-2026

[13] Şahin S, Samli R, Tan AS, Barba FJ, Chemat F, Cravotto G, et al. Solvent-free microwave-assisted extraction of polyphenols from olive tree leaves: Antioxidant and antimicrobial properties. Molecules. 2017;**7**:1054

[14] Wang J, Sun B, Cao Y, Tian Y, Li X. Optimisation of ultrasound-assisted extraction of phenolic compounds from wheat bran. Food Chemistry. 2008;**106**: 804-810

[15] Wong F. Slope stability and response surface method. Journal of Geotechnical and Geo-environmental Engineering ASCE. 1985;**111**:32-53

[16] Tandjiria VI, Teh CI, Low BK. Reliability analysis of laterally loaded piles using response surface methods. Structural Safety. 2000;**22**:335-355

[17] Sivakumar BG, Srivastava A. Response Surface Methodology (RSM)

in the Reliability Analysis of Geotechnical Systems. In: Proceedings of 12th International conference of International Association for Computer Methods and Advances in Geomechanics. Goa India; 2008

[18] Wan Nor Nadyaini Wan Omar. Design of Experiment (DOE) and Response Surface Methodology (RSM). N29 Faculty of Chemical Engineering; 2015

[19] Minitab's 20 Help Section and DOE Navigation.

[20] Introduction to Design of Experiment [Internet]. 2019. Available from: www.theopeneducator.com

[21] Behera SK, Meena H, Chakraborty S, Meikap BC. Application of response surface methodology (RSM) for optimization of leaching parameters for ash reduction from low-grade coal. International Journal of Mining Science and Technology. 2018;**28**:621-629

[22] Tian Y, Xu Z, Zheng B, Martin LY. Optimization of ultrasonic-assisted extraction of pomegranate (Punica granatum L.) seed oil. Ultrasonics Sonochemistry. 2013;**20**:202-208

[23] Samaram S, Mirhosseini H, Tan CP, Ghazali HM, Bordbar S, Serjouie A. Optimisation of ultrasound-assisted extraction of oil from papaya seed by response surface methodology: Oil recovery, radical scavenging antioxidant activity, and oxidation stability. Food Chemistry. 2015;**172**:7-17

[24] Cornell J. How to Apply Response Surface Methodology. American Society for Quality Control. USA: Statistic Division; 1990

[25] Brown SR, Melemend LE. Experimental Design and Analysis Quantitative Application in the Social Science. California: Sage Publication; 1990. p. 74

[26] Fractional factorial design [Internet]. Available from: http://en.wikipedia.org/wiki/Fractional_factorial_design#Resolution.

[27] Raliasoft DOE Design of Experiment techniques [Internet]. 1992. Available from: www.help.synthesisplatform.net

[28] Neto A, Oliveira S. A Software Tools Catalogue to Support the Statistical Process Control on the Software Context. In: Proceedings of the 14th International Conference on Evaluation of Novel Approaches to Software Engineering. 2019. pp. 510-517

[29] Paul G. Mathews Design of Experiments with MINITAB. Milwaukee, Wisconsin: ASQ Quality Press;

[30] Kumari M, Gupta SK. Response surface methodological (RSM) approach for optimizing the removal of trihalomethanes (THMs) and its precursor's by surfactant modified magnetic nanoadsorbents (sMNP) – An endeavour to diminish probable cancer risk. Scientific Reports. 2019;**9**:18339

[31] Dogan IS, Okut H. As a Tool Response Surface Methodology (RSM) in a New Product Development. In: International congress on Information Technology in Agriculture, Food and Environmental. Izmir, Turkey: Ege University; 2003

[32] Russell V. Surface Plots in the RSM Package Updated to version 2.00. Iowa: Department of Statistics, The University of Iowa; 2012

[33] Box GEP, Hunter WG, Hunter JS. Statistics for Experimenters. An

Introduction to Design, Data Analysis, and Model Building. 2nd ed. New York: John Wiley & Sons; 2005

[34] Noelle M.. How to Use MINITAB, Design of Experiments, 2014.

[35] Khan RM. Problem-solving and Data Analysis Using Minitab. A Clear and Easy Guide to Six Sigma Methodology. 1st ed. West Sussex: Wiley; 2013

Chapter 4

Perspective Chapter: Cyclic Generation of Box-Behnken Designs and New Second-Order Designs

Nam-Ky Nguyen, John J. Borkowski and Mai Phuong Vuong

Abstract

Box-Behnken designs (BBDs) are three-level second-order spherical designs with all points lying on a sphere, introduced by Box and Behnken, for fitting the second-order response surface models. They are available for 3–12 and 16 factors. Together with the central composite designs for the second-order model, BBDs are very popular response surface designs, especially for 3–7 factors. This chapter introduces an algorithm to produce cyclic generators for BBDs and similar designs, which we call cyclic BBDs (CBBDs). The new CBBDs offer more flexibility in choosing the designs for a specified number of factors. Comparisons between some BBDs and the new CBBDs indicate the superiority of the new CBBDs with respect to multiple design quality measures and graphical tools assessing prediction variance properties. A catalog of 24 new CBBDs, which includes orthogonally blocked CBBDs for 11, 13, and 14 factors, will be given.

Keywords: circulant matrices, foldover designs, interchange algorithm, response surface designs, spherical designs

1. Introduction

Box-Behnken designs (BBDs) are three-level response surface designs (RSDs), introduced by Box and Behnken [1, 2], to fit a second-order response surface model

$$\mathbf{y} = \mathbf{X}\beta + \varepsilon \tag{1}$$

For m factors in n runs. Here, $\mathbf{y}_{n\times 1}$ is a response vector; $\mathbf{X}_{n\times p}$ the model matrix having an intercept term, m main effect (ME) terms, m quadratic effect (QE) terms, and $\binom{m}{2}$ 2-factor interaction (2FI) terms; vector $\beta_{p\times 1}$ of $p = 1 + 2m + \binom{m}{2}$ parameters; and error vector $\varepsilon_{n\times 1}$ with zero mean and covariance matrix $\mathbf{I}\sigma^2$. BBDs are currently available for 3–12 and 16 factors [3]. Except for 11 factors, BBDs can be constructed by superimposing the two-level factorial design onto treatments in each block of a balanced incomplete block design (IBD) or partially balanced IBD. BBDs have the following properties:

i. Each factor has the same number of runs at high (+1) and low (−1) levels;

IntechOpen

ii. All points lie on a sphere of the radius ρ or at the center of the design space;

iii. They are rotatable for 4 and 7 factors. Otherwise, they are near-rotatable;

iv. They can be orthogonally blocked except for 3 and 11 factors;

v. Let design D be a $n \times m$ design matrix D with m factors $x_1, \dots, x_m$. Let the row u of the model matrix $\mathbf{X}$ be written as $\left(1, x_{u1}^2, \dots, x_{um}^2, x_{u1}, \dots, x_{um}, x_{u1}x_{u2}, \dots, x_{u(m-1)}x_{um}\right)$, where x_{ui} is the element in row u and column i of D. The information matrix $\mathbf{M} = \mathbf{X}'\mathbf{X}$ (and its inverse) has the following form:

$$\left(\begin{array}{c|c} \mathbf{M}_{11} & \mathbf{M}_{12} \\ \hline \mathbf{M}_{21} & \mathbf{M}_{22} \end{array}\right), \tag{2}$$

where $\mathbf{M}_{11}$ is a square matrix of order $1 + m$, and $\mathbf{M}_{22}$ is a square matrix of order $m + \binom{m}{2}$. For a BBD, $\mathbf{M}_{21} = \mathbf{0}$, $\mathbf{M}_{12} = \mathbf{0}'$ and $\mathbf{M}_{22} = \mathbf{D}$, where $\mathbf{D}$ is a diagonal matrix. Matrix $\mathbf{M}$ in (2) reduces to:

$$\left(\begin{array}{c|c} \mathbf{M}_{11} & \mathbf{0}' \\ \hline \mathbf{0} & \mathbf{D} \end{array}\right). \tag{3}$$

As an example, we construct a 6-factor BBD. Consider an IBD of size (v, k, r)= (6, 3, 4) for six varieties, arranged in blocks of size three, each with three replications per variety. Superimposing a 2^3 factorial onto the corresponding varieties of this IBD will result in the following 6-factor BBD without center points:

$$\begin{pmatrix} \pm1 & \pm1 & 0 & \pm1 & 0 & 0 \\ 0 & \pm1 & \pm1 & 0 & \pm1 & 0 \\ 0 & 0 & \pm1 & \pm1 & 0 & \pm1 \\ \pm1 & 0 & \pm1 & \pm1 & 0 & 0 \\ 0 & \pm1 & 0 & 0 & \pm1 & \pm1 \\ \pm1 & 0 & \pm1 & 0 & 0 & \pm1 \end{pmatrix}.$$

In each row, $(\pm1 \pm 1 \pm 1)$ represents the eight points of a 2^3 design and 0 is a column vector of eight 0's. Czyrski and Sznura [4] applied the 6-factor BBD in the optimization of HPLC separation of fluoroquinolones.

Next, we examine a foldover design in 48 runs (with no center points) generated by four cyclic generators: $(-1, 0, 0, -1, 1, 0)$, $(0, 1, 0, 0, 1, 1)$, $(0, 0, 1, -1, 0, -1)$, and $(0, 0, -1, -1, 0, 1)$. The first generator, for example, cyclically generates six design points:

$$\begin{pmatrix} -1 & 0 & 0 & -1 & 1 & 0 \\ 0 & -1 & 0 & 0 & -1 & 1 \\ 1 & 0 & -1 & 0 & 0 & -1 \\ -1 & 1 & 0 & -1 & 0 & 0 \\ 0 & -1 & 1 & 0 & -1 & 0 \\ 0 & 0 & -1 & 1 & 0 & -1 \end{pmatrix}.$$

The four cyclic generators produce 24 runs. The next 24 runs are obtained by folding over the first 24 runs (i.e., changing the signs of the factor levels). All points lie on a sphere of radius $\rho = \sqrt{3}$. It can be shown that these design points are also points in the 6-factor BBD. In this chapter, we call this type of design a cyclic BBD or CBBD.

Each factor of this BBD has half of its runs at the 0-level and the remaining at ± 1 levels. Now assume that the researchers are looking for an alternative spherical design with fewer 0-levels and more ± 1 levels for each factor. This allows the experimenter to increase the volume of the spherical design region by increasing the radius associated with CBBD points. This chapter introduces an algorithm that can generate CBBDs of varying radii. Designs with the same number of factors and runs but with different radii are compared with respect to D-criterion values (or d-values), variances of the parameter estimates, and the correlation among the main (ME), quadratic (QE), and interaction (2FI) effects. Concepts, such as rotatability, orthogonal blocking, and spherical designs, are well-described in Box and Behnken [2] and textbooks on response surface methodology, such as Myers et al. [5] or Box and Draper [6].

2. Calculating the elements of M of a CBBD

The design matrix D of a CBBD has the form $\left(C_1' \ldots C_r'\ 0'\right)'$ where $C_1, \ldots, C_r$ are the circulant matrices of order m generated by r generating vectors $c_1, \ldots, c_r$ and 0 is a matrix containing center points. For the information matrix **M** to have the form in (3), the elements D must satisfy the following conditions:

$$\sum_{u=1}^{n} x_{ui} = 0\ (\forall i) \tag{4}$$

$$\sum_{u=1}^{n} x_{ui}x_{uj} = 0\ (i \neq j) \tag{5}$$

$$\sum_{u=1}^{n} x_{ui}^2 x_{uj} = 0\ (i \neq j) \tag{6}$$

$$\sum_{u=1}^{n} x_{ui}^2 x_{uj} x_{uk} = 0\ (i \neq j \neq k) \tag{7}$$

$$\sum_{u=1}^{n} x_{ui} x_{uj} x_{uk} = 0\ (i \neq j \neq k) \tag{8}$$

$$\sum_{u=1}^{n} x_{ui} x_{uj} x_{uk} x_{ul} = 0\ (i \neq j \neq k \neq l) \tag{9}$$

where x_{ui} is the level of the factor i for run u (Cf. Appendix A of [2]). The condition in (4) implies that D is a balanced design; that is, each column of D has the same number of $+1$ and -1. To make D balanced, we just have to restrict the sum of the elements of the generating vectors $c_1, \ldots, c_r$ to 0. As D is constructed from the circulant matrices, conditions (5)–(9) can be written as:

$$\sum_{t=1}^{r} \sum_{i=1}^{m-1} c_{ti} c_{t(i+j) \bmod m} = 0\ (1 \leq j < m) \tag{10}$$

$$\sum_{t=1}^{r}\sum_{i=1}^{m-1} c_{ti}^2 c_{t(i+j)\text{mod } m} = 0\ (1 \le j < m) \tag{11}$$

$$\sum_{t=1}^{r}\sum_{i=1}^{m-1} c_{ti}^2 c_{t(i+j)\text{mod } m} c_{t(i+k)\text{mod } m} = 0\ (1 \le j < k < m) \tag{12}$$

$$\sum_{t=1}^{r}\sum_{i=1}^{m-1} c_{ti} c_{t(i+j)\text{mod } m} c_{t(i+k)\text{mod } m} = 0\ (1 \le j < k < m) \tag{13}$$

$$\sum_{t=1}^{r}\sum_{i=1}^{m-1} c_{ti} c_{t(i+j)\text{mod } m} c_{t(i+k)\text{mod } m} c_{t(i+l)\text{mod } m} = 0\ (1 \le j < k < l < m) \tag{14}$$

where c_{ti} is the value of the factor i on the generating vector t. It can be seen that there are $m-1$ summations in (10) and (11), $\binom{m-1}{2}$ in (12) and (13), and $\binom{m-1}{3}$ in (14). This explains why the lengths of the vectors J_q and J in Section 3 are $2(m-1)+2\binom{m-1}{2}+\binom{m-1}{3}$.

3. The CBBD algorithm

Our CBBD algorithm is the generalization of the algorithm in Nguyen et al. [7] and Pham et al. [8]. Using the results in Section 2, we present the steps of the algorithm for generating a CBBD for m factors in $n = 2rm + n_c$ runs (where n_c is the number of center points) with points on a sphere of radius ρ, and $\left(\frac{1}{3}m \le \rho^2 < m\right)$.

1. Form a matrix C of size $r \times m$. Set $\frac{1}{2}r\rho^2$ elements of C to 1, $\frac{1}{2}r\rho^2$ to -1, and the remaining elements to 0. For each row vector c_q of C, form a vector J_q with a length $2(m-1)+2\binom{m-1}{2}+\binom{m-1}{3}$ containing the sums in (10) to (14). Set $J = \sum J_q$. Calculate f, the sum of the squares of the elements of J.

2. Search for a pair of entries in C such that swapping their positions results in the biggest reduction in f. If the search is successful, update f and C. This step is repeated until $f = 0$, or f cannot be reduced further.

Remarks

1. These two steps make up one trial. Among all trials with $f = 0$, we select the CBBD with the highest D-criterion value, which is defined as:

$$d\text{-value} = \frac{1}{n}|\mathbf{M}|^{1/p} \tag{15}$$

   for the information matrix $\mathbf{M}$ and the number of parameters p for the second-order model.

2. There are situations, where there is no CBBD with $f = 0$ for particular values of m, ρ^2 and r. In this case, we compute two values f_1 and f_2, set f_1 equal to the sum of squares of the first $2(m-1)+2\binom{m-1}{2}$ elements of J (or the first $2(m-1)+$

$\binom{m-1}{2}$ elements of J) and f_2 the sum of squares of the remaining elements. A design is selected if $f_1 = 0, f_2$ is minimum and the d-value in (15) is maximum.

3. If D is a foldover design, the sums in Eqs. (11) and (13) will be 0, and the length of the vector J_q and J is shortened to $(m-1) + \binom{m-1}{2} + \binom{m-1}{3}$.

4. BBDs and new CBBDs

Table 1 displays the quality measures of BBDs whose run sizes (excluding the two center runs) are multiples of the number of factors m and 24 CBBDs. **Table 1** does not include two BBDs for (m, ρ^2) = (9, 3) and (16,4) due to their over-abundance of 0-factor levels. This table includes m (the number of factors), ρ^2 (the square of the radius), n (the run size of each BBD which includes two center points), and the quality measures of the designs. These measures are the d-value in (15), v_Q, v_M, and v_I (the maximum scaled variances of the QEs, MEs, and 2FIs, respectively), r_{QQ}, r_{QI}, r_{MI}, and r_{II} (the maximum of the absolute values of the correlations between two QEs, between a QE and a 2-FI, between a ME and a 2FI, and between 2FIs, respectively). Note that r_{QM} (the correlation between a QE and a ME) and r_{MM} (the correlation between two MEs) for all designs in **Table 1** are always zero.

Out of 24 CBBDs in **Table 1**, there are 15 CBBDs with $f = 0$ using the foldover technique with the first half-fraction being balanced with factors having the same number of ± 1's. The first half-fraction of the CBBDs for 3–7 and for 8–14 factors in this table require four and eight cyclic generators, respectively. Like BBDs, these CBBDs have $r_{QI} = r_{MI} = r_{II} = 0$. Also, like BBDs, they can be orthogonally blocked, with each half-fraction forming a block. The four CBBDs that are identical to BBDs in terms of quality measures are the ones for 5, 6, 7, and 12 factors. Note that for 3 and 4 factors, the CBBDs have more runs than the corresponding BBDs, and, hence, provide more error degrees of freedom. Also, the 8-factor BBD requires many more runs (nearly 200) than the CBBD. The BBD for 11 factors cannot be orthogonally blocked, and BBDs for 13 and 14 factors are not available. It is necessary to mention that the designs in Nguyen and Borkowski [9] are not the foldover CBBDs in **Table 1**, and as such cannot be blocked in the same way.

There are nine CBBDs for 3–8 factors that are constructed without applying the foldover technique to the first half-fraction. We denote these CBBDs as CBBD*s. The CBBD* for three factors requires four cyclic generators, while all others require eight. CBBD*s for 5–8 factors have $f_1 = 0$ (see Remark 2 of Section 3). These designs cannot be blocked in the same way as the CBBDs in **Table 1**. They can, however, be nearly orthogonally blocked using suitable software (see [10]).

These CBBDs and CBBD*s offer additional design choices to an experimenter. Comparisons of CBBDs and CCBD*s to the BBDs for the same number of factors and runs indicate that they, in general, have higher d-values, smaller variances of the estimates, and smaller r_{QQ} (the correlation between two different quadratic effects). **Figure 1** displays the color cell plots (CCPs) of BBDs for 5–8 factors, that is, 5a, 6a, 7a, and 8a, and the corresponding CBBD*s with $\rho^2 = m - 1$, that is, 5c, 6b, 7d, and 8f. CCPs, proposed by Jones and Nachtsheim [11], display the magnitude of the correlation between the columns of the model matrix **X** (in terms of the absolute values). The color of each cell ranges from white (no or near-zero correlation) to dark (one or near-one correlation). It can be seen from these CCPs that the information matrices **M** of

Design	#	m	ρ^2	n	d-value	v_Q	v_M	v_I	r_{QQ}	r_{QI}	r_{MI}	r_{II}
BBD/CBBD*	3a	3	2	14	.377	.313	.125	.250	.167	0	0	0
CBBD	3b		2	26	.379	.219	.063	.125	.300	0	0	0
BBD	4a	4	2	26	.255	.229	.083	.250	.238	0	0	0
CBBD	4b		2	34	.246	.211	.063	.250	.417	0	0	0
CBBD	4c		3	34	.439	.153	.042	.063	.133	0	0	0
BBD/CBBD	5a	5	2	42	.174	.198	.063	.250	.212	0	0	0
CBBD	5b		3	42	.303	.208	.042	.125	.556	0	0	0
CBBD*	5c		4	42	.429	.133	.031	.068	.050	0	0	.333
BBD/CBBD	6a	6	3	50	.243	.134	.042	.125	.359	0	0	0
CBBD*	6b		5	50	.484	.125	.025	.039	0	0	0	.25
BBD/CBBD	7a	7	3	58	.196	.111	.042	.125	.137	0	0	0
CBBD*	7b		4	58	.276	.086	.031	.094	.115	0	0	.500
CBBD*	7c		5	58	.370	.235	.025	.071	.356	0	0	.333
CBBD*	7d		6	58	.516	.122	.021	.028	.033	0	0	.200
BBD‡	8a	8	3	194	.150	.075	.014	.063	.237	0	0	0
CBBD	8b		3	130	.148	.085	.021	.125	.321	0	0	0
CBBD	8c		4	130	.251	.057	.016	.042	.231	0	0	0
CBBD*	8d		3	66	.124	.115	.130	.380	.310	0	.408	0
CBBD*	8e		4	66	.225	.083	.058	.109	.213	0	.204	0
CBBD*	8f		7	66	.454	.120	.032	.044	.057	0	.154	.167
BBD‡	9a		4	146	.184	.160	.016	.063	.335	0	0	0
CBBD	9b		4	146	.194	.063	.016	.063	.335	0	0	0
BBD	10a	10	4	162	.160	.064	.016	.063	.240	0	0	0
CBBD	10b		4	162	.166	.055	.016	.063	.240	0	0	0
BBD	11a	11	5	178	.215	.039	.013	.031	.090	0	0	0
CBBD	11b		4	178	.136	.058	.016	.063	.219	0	0	0
BBD/CBBD	12	12	4	194	.118	.054	.016	.063	.254	0	0	0
CBBD	13	13	4	210	.103	.051	.016	.063	.079	0	0	0
CBBD	14	14	4	226	.083	.054	.016	.125	.302	0	0	0

*†Each design run size n includes two center runs. All BBDs can be orthogonally blocked except BBDs for $m = 3, 11$ factors (3a and 11a). CBBDs require $r = (n-2)/2m$ cyclic generators. CBBD*s require $r = (n-2)/m$ cyclic generators.*
‡The two BBDs for $m = 8, 9$ (8a and 9a) appear in Box and Behnken [1].

Table 1.
*Quality measures of BBDs, CBBDs, and CBBD*s,†.*

the mentioned CBBD*s do not have the form in (3), but all QEs are orthogonal to all MEs and 2FIs. Note that the BBD for 8 factors has 194 runs, while the corresponding CBBD* has only 66 runs.

Appendices A and B display the cyclic generators of the CBBDs and CBBD*s respectively, in **Table 1**.

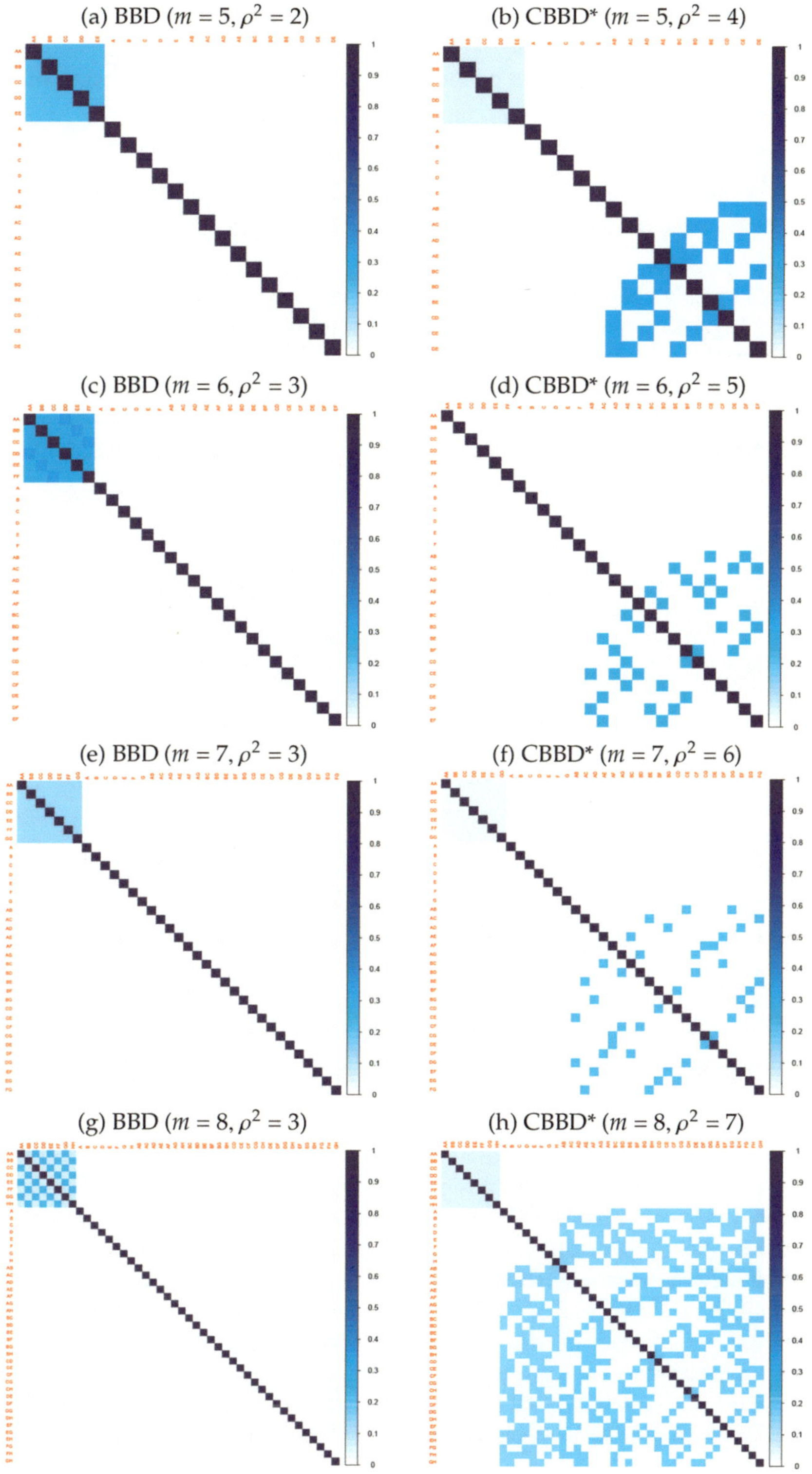

Figure 1.
*CCPs for BBDs and CBBD*s with $\rho^2 = m - 1$ ($m = 5,6,7,8$).*

5. FDS plot and VDG comparisons

When assessing the prediction properties of an RSD, we want a design that will produce predicted values $\hat{Y}_{(x_1, \dots, x_m)}$ with low variance for points $(x_1, \dots, x_m)$ in the design space. The prediction variance at $(x_1, \dots, x_m)$ is $\text{var}\left(\hat{Y}_{(x_1, \dots, x_m)}\right) = \sigma^2 \mathbf{x}(\mathbf{X}'\mathbf{X})^{-1}\mathbf{x}'$, where σ^2 is the error variance and $\mathbf{x}$ is $(x_1, \dots, x_m)$ expanded to contain the m_2 second-order model terms. Re-scaling by n/σ^2 yields the scaled prediction variance $V(x_1, \dots, x_m) = n\mathbf{x}(\mathbf{X}'\mathbf{X})^{-1}\mathbf{x}'$.

Although a design efficiency measure (such as the d-value) may provide useful information regarding the overall quality of prediction, it does not provide information regarding the distribution of the prediction variance throughout the design region. This is addressed by studying a design's spherical prediction variance (SPV) properties.

V_ρ is defined to be the *average of the scaled prediction variance* function taken over S_ρ, the sphere of radius ρ. (See [12]) Thus,

$$V_\rho = \frac{1}{\omega_\rho}\int_{S_\rho} V(x_1, \dots, x_m)\ dx_1 \dots dx_m \tag{16}$$

where ω_ρ is the surface area of S_ρ. Also of interest are the *minimum* and *maximum scaled prediction variances* defined as:

$$VMIN_\rho = \min_{(x_1, \dots, x_m) \in S_\rho} V(x_1, \dots, x_m) \tag{17}$$

$$VMAX_\rho = \min_{(x_1, \dots, x_m) \in S_\rho} V(x_1, \dots, x_m) \tag{18}$$

Fraction of design space (FDS) plots and variance dispersion graphs (VDGs) will be utilized to assess the prediction variance properties of designs in **Table 1**. Giovannitti-Jensen and Myers [13] introduced the VDG, which superimposes plots of $VMAX_\rho$, $VMIN_\rho$, and V_ρ against the radius ρ within a spherical design space. Modified VDGs that also include the SPV values of $V(x_1, \dots, x_m)$ for a large set of random points in the spherical region [9] will be presented. Note that the proportion of the volume of the design region is small for values of ρ near-zero but increases rapidly with increase ρ. Thus, a large proportion of the design space is associated with a relatively small interval ρ near the design space boundary. To address this issue, Zahran et al. [14] introduced the FDS plot of the quantiles of $V(x_1, \dots, x_m)$ against the fraction (or proportion) of the volume of the design region. Unlike single-valued design efficiency measures, both VDGs and FDS plots allow a more thorough assessment throughout the design region. For a summary of graphical methods for assessing the prediction variance properties of RSDs, see Borkowski [15].

Before a comparison of designs using these graphical tools can be made, a critical issue involving factor scaling needs to be addressed. A major difficulty in comparing a BBD to a CBBD or CBBD* with the same design size n is that the design spaces are not the same. For example, consider the BBD with $(m, \rho^2) = (5, 2)$, that is, 5a. Calculation of v_Q, v_M, and v_I is based on the assumption that the design region includes points within the 5-dimensional hypersphere of radius $\sqrt{2}$. However, for the CBBD* with $(m, \rho^2) = (5, 4)$, that is, 5c, the calculation of v_Q, v_M, and v_I are based on the assumption that the design region includes points within the 5-dimensional hypersphere of radius $\sqrt{4}$.

Consider the following five-factor experiment presented in Myers et al. [5]. The response to be analyzed is rayon whiteness (RW), which is associated with fabric quality. The experimenters believed that RW can be affected by process variables, which include acid bath temperature in °C (temp1), percent acid concentration (conc1), water temperature in °C (temp2), sulfide concentration (conc2), and amount of chlorine bleach in lb./min (bleach). The experimental levels and the coded levels x_1, x_2, x_3, x_4, x_5 for the five variables are as follows:

Coded			**Experimental levels**		
Levels	**temp1 (°C)**	**conc1 (%)**	**temp2 (°C)**	**conc2 (%)**	**bleach (lb/min)**
−1	35	.3	82	.20	.3
0	45	.5	85	.25	.4
1	55	.7	88	.30	.5

Table 2 shows the 42 design points for the BBD with $\rho^2 = 2$, the CBBD with $\rho^2 = 3$, and the CBBD* with $\rho^2 = 4$ (designs 5a, 5b, and 5c, respectively). If any 0-factor level is replaced with a value > 0 or < 0 in any of these designs, then that point is outside that experiment's design space. There is an important implicit assumption that the fitted model will be appropriate when extrapolating outside the design space. This can be dangerous because it can not only result in predictions with increased bias but also result in larger prediction variances. Whether or not bias is introduced when extrapolating, increasing variances will occur and can be seen in the comparison of VDGs.

Therefore, to make comparisons between designs 5a, 5b, and 5c when choosing a design, it is reasonable to assume that the coded factor levels of $-1, 0, 1$ representing the same levels when uncoded. This will be true for all design comparisons made for $m = 3, \dots, 11$ factors in **Table 1**.

We begin our comparisons between designs 5a, 5b, and 5c by generating FDS plots and VDGs over the maximum ρ^2, which are seen in **Figure 2**. For $m = 5$, that would be $\rho^2 = 4$. In the VDGs, vertical reference lines are placed at $\rho = \sqrt{2}$ and $\rho = \sqrt{3}$, which represent the maximum ρ for points in designs 5a and 5b, respectively. The FDS plots are based on the distribution of the SPV values for 10,000 randomly selected points in a sphere of radius $\sqrt{4}$. The 10,000 ($m = 4, 5, 6, 7$) or 20,000 ($m = 8$) SPV values are also plotted in the VDGs (as suggested in [9]).

To compare the five-factor designs, the VDGs in **Figure 2** should be examined over three disjoint intervals for the radius: (i) $[0, \sqrt{2}]$, (ii) $[\sqrt{2}, \sqrt{3}]$, and (iii) $[\sqrt{3}, \sqrt{4}]$. For (i), the maximum and average SPV is best for the BBD followed by the CBBD* 5c and CBBD 5b. This should not be surprising because every BBD design point is within $\sqrt{2}$ of the origin. However, for (ii) and (iii), it is clear that the CBBD* is best for having smaller maximum, average, and minimum SPV values over $\rho \in [\sqrt{2}, \sqrt{4}]$. These plots indicate that the BBD is best only if the experimenter does not plan to predict the mean response at points with $\rho > \sqrt{2}$ (such as at ($\pm 1, \pm 1, \pm 1, 0, 0$) or ($\pm 1, \pm 1, \pm 1, \pm 1, 0$)). This seems unrealistic. As stated earlier, if any 0-factor level is changed, then the negative consequences of extrapolation must be acknowledged.

BBD ($\rho^2 = 2$)					CBBD ($\rho^2 = 3$)					CBBD* ($\rho^2 = 4$)				
x_1	x_2	x_3	x_4	x_5	x_1	x_2	x_3	x_4	x_5	x_1	x_2	x_3	x_4	x_5
0	0	1	0	1	−1	0	0	−1	1	1	1	0	1	1
1	0	0	1	0	1	−1	0	0	−1	1	1	1	0	1
0	1	0	0	1	−1	1	−1	0	0	1	1	1	1	0
1	0	1	0	0	0	−1	1	−1	0	0	1	1	1	1
0	1	0	1	0	0	0	−1	1	−1	1	0	1	1	1
0	−1	−1	0	0	1	−1	−1	0	0	−1	1	−1	0	1
0	0	−1	−1	0	0	1	−1	−1	0	1	−1	1	−1	0
0	0	0	−1	−1	0	0	1	−1	−1	0	1	−1	1	−1
−1	0	0	0	−1	−1	0	0	1	−1	−1	0	1	−1	1
−1	−1	0	0	0	−1	−1	0	0	1	1	−1	0	1	−1
1	0	0	−1	0	0	0	1	1	1	0	−1	−1	−1	−1
0	1	0	0	−1	1	0	0	1	1	−1	0	−1	−1	−1
−1	0	1	0	0	1	1	0	0	1	−1	−1	0	−1	−1
0	−1	0	1	0	1	1	1	0	0	−1	−1	−1	0	−1
0	0	−1	0	1	0	1	1	1	0	−1	−1	−1	−1	0
1	−1	0	0	0	−1	1	0	0	−1	−1	0	−1	1	1
0	1	−1	0	0	−1	−1	1	0	0	1	−1	0	−1	1
0	0	1	−1	0	0	−1	−1	1	0	1	1	−1	0	−1
0	0	0	1	−1	0	0	−1	−1	1	−1	1	1	−1	0
−1	0	0	0	1	1	0	0	−1	−1	0	−1	1	1	−1
0	0	−1	0	−1	1	0	0	1	−1	−1	1	0	1	−1
−1	0	0	−1	0	−1	1	0	0	1	−1	−1	1	0	1
0	−1	0	0	−1	1	−1	1	0	0	1	−1	−1	1	0
−1	0	−1	0	0	0	1	−1	1	0	0	1	−1	−1	1
0	−1	0	−1	0	0	0	1	−1	1	1	0	1	−1	−1
0	1	1	0	0	−1	1	1	0	0	−1	1	1	0	−1
0	0	1	1	0	0	−1	1	1	0	−1	−1	1	1	0
0	0	0	1	1	0	0	−1	1	1	0	−1	−1	1	1
1	0	0	0	1	1	0	0	−1	1	1	0	−1	−1	1
1	1	0	0	0	1	1	0	0	−1	1	1	0	−1	−1
−1	0	0	1	0	0	0	−1	−1	−1	1	0	−1	1	−1
0	−1	0	0	1	−1	0	0	−1	−1	−1	1	0	−1	1
1	0	−1	0	0	−1	−1	0	0	−1	1	−1	1	0	−1
0	1	0	−1	0	−1	−1	−1	0	0	−1	1	−1	1	0
0	0	1	0	−1	0	−1	−1	−1	0	0	−1	1	−1	1
−1	1	0	0	0	1	−1	0	0	1	0	1	1	−1	−1
0	−1	1	0	0	1	1	−1	0	0	−1	0	1	1	−1
0	0	−1	1	0	0	1	1	−1	0	−1	−1	0	1	1
0	0	0	−1	1	0	0	1	1	−1	1	−1	−1	0	1
1	0	0	0	−1	−1	0	0	1	1	1	1	−1	−1	0
0	0	0	0	0	0	0	0	0	0	0	0	0	0	0
0	0	0	0	0	0	0	0	0	0	0	0	0	0	0

The first point (row) in each circulant block of five points generates the other four points cyclically. For the BBD and CBBD, the first 20 points are folded over to form the second 20 points. Each design has two center points to form these 42-point designs.

Table 2.
Design points for the five-factor BBD, CBBD, and CBBD.*

Predictions based on the BBD at such points are extrapolations leading to larger SPV values. This is reflected in $(v_Q, v_M, v_I) = (.198, .063, .250)$ for the BBD and $(.031, .068, .050)$ for the CBBD*. These values indicate that the estimated

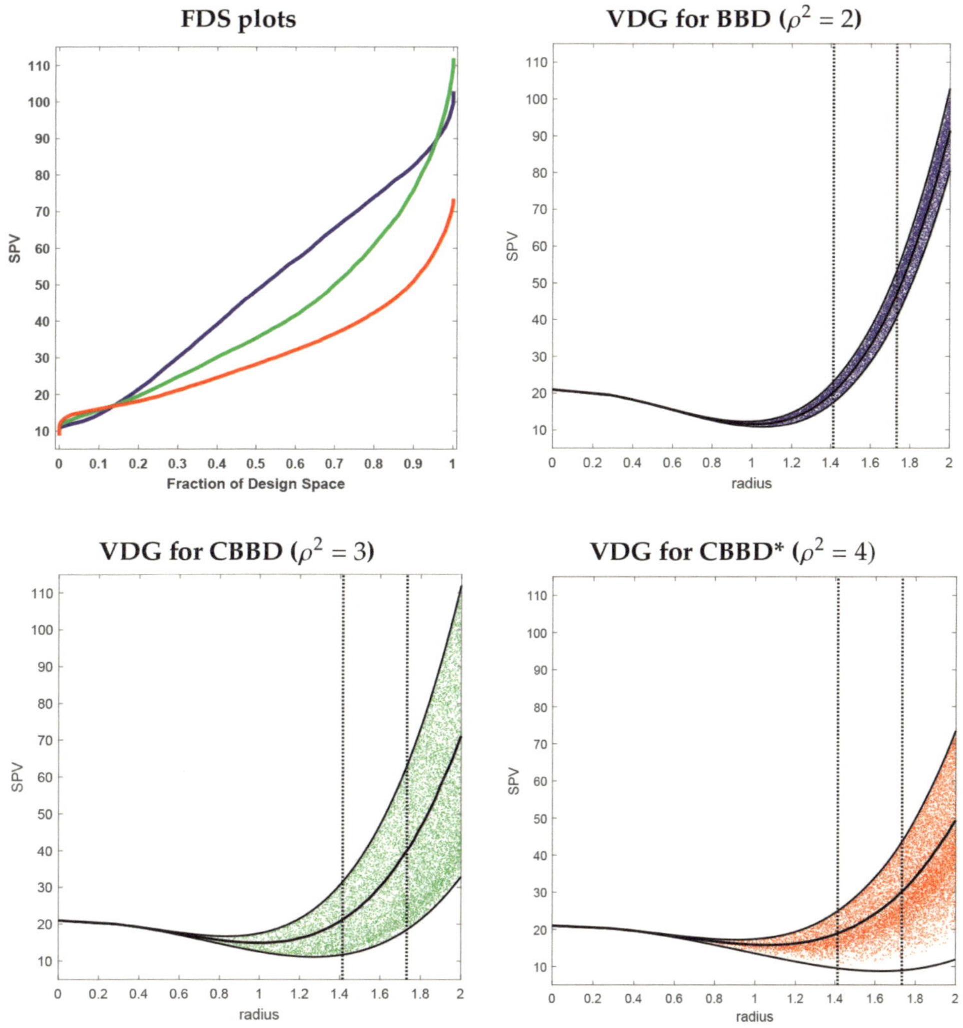

Figure 2.
FDS plots and VDGs for designs with 5 factors ($n = 42$). FDS lines: blue for BBD, green for CBBD, and red for CBBD. VDGs include solid black lines for the minimum, average, and maximum SPV. Vertical reference lines are plotted at $\sqrt{2}$ and $\sqrt{3}$.*

parameter variances associated with the CBBD* are smaller than those for the BBD for $\rho \in [\sqrt{2}, \sqrt{4}]$. Thus, we would expect better predictions with the CBBD*. This is supported by the VDGs and the CBBD* having the largest d-value. The CBBD is the least desirable of the $m = 5$ factor designs primarily due to the large $v_Q = .208$ value.

Using the comparison approach applied to the five-factor designs, we now summarize the comparison of equal-sized designs for $m = 4,6,7$, and 8 factors.

For the four-factor designs with $n = 34$, the FDS plot and VDG in column 1 of **Figure 3** for the CBBD with $\rho^2 = 3$, that is, 4c, are superior to the BBD with $\rho^2 = 2$, that is, 4b, especially over the interval $[\sqrt{2}, \sqrt{3}]$, where it has the smaller maximum, average, and minimum SPV values. This is expected because no extrapolation occurs over this interval for 4c, while it does for 4b. These plots indicate that design 4b is best only if the experimenter does not plan to predict the mean response at points with $\rho > \sqrt{2}$. This is reflected in the larger d-value and smaller (v_Q, v_M, v_I) for design 4b.

For the six-factor designs with $n = 50$, the FDS plot and VDG in column 2 of **Figure 3** for the CBBD* with $\rho^2 = 5$ (design 6b) are superior to the BBD/CBBD with $\rho^2 = 3$ (design 6a) for most of the design space. The only exception is for a small fraction of the design space, where ρ^2 is close to $\sqrt{5}$ and maximum SPV values are larger for design 6b. Despite this small subregion, design 6b has the smaller average and minimum SPV values over the interval $\left[\sqrt{3}, \sqrt{5}\right]_,$ which comprises most of the spherical volume. Design 6b also has a larger d-value and smaller (v_Q, v_M, v_I).

For seven factors, there are four designs with $n = 58$. The FDS plots in **Figure 4** indicate that the CBBD* with $\rho^2 = 4$, that is, 7b, is the best design over a spherical design space of radius $\sqrt{6}$. The VDGs also indicate that this design has the smallest maximum, average, and minimum SPV values for $\rho > \sqrt{3}$, and based on the concentration of SPV values near the maximum for any ρ, the distribution of SPV values is highly skewed-left. The experimenter, however, must realize that beyond $\rho > \sqrt{4}$, extrapolation occurs for 7b and the experimenter is ignoring the possibility that increased bias may exist with predictions when using the fitted model that results from the experimental data. The VDG and FDS plot for 7c indicates that the geometry of the design points in the design space is poor despite having $\rho^2 = 5$. This indicates that in certain cases, a design with a larger ρ^2 value does not necessarily guarantee a better design. It is important to note, however, that this case is a rare exception. The BBD/CBBD 7a is rotatable. Therefore, the minimum, maximum, and average SPVs are all equal for a given radius. This is reflected in the single curve in its VDG. The VDG for the $\rho^2 = 5$ CBBD* is truncated at SPV = 240 for scale clarity when making VDG comparisons.

For eight factors, there are three CBBB* designs with $n = 66$. The FDS plots in **Figure 5** suggest that the CBBD* with $\rho^2 = 4$ (design 8e) is the best design over a spherical design space of radius $\sqrt{7}$. The VDGs also indicate that this design has the smallest maximum and average SPV values for $\rho > \sqrt{3}$. It is important to remind the experimenter that between $\rho = \sqrt{4}$ and $\sqrt{7}$, extrapolation is occurring for 8e. Thus, although 8e appears better than 8f, there may be increased bias with any prediction associated with using a fitted model for 8e in comparison with 8f over this interval. Note that although the minimum SPV curve for $\rho^2 = 7$ CBBD* (design 8f) is the lowest for $\rho > \sqrt{3}$, it is associated with only a small fraction of the design space as evidenced by the sparsity of points near the minimum. The VDG for the $\rho^2 = 3$ CBBD* is truncated at SPV = 375 for scale clarity when making VDG comparisons.

Based on the comparisons for $m = 7$and 8 factors, the design with the largest d-value is not necessarily the best design when using FDS plots and VDGs as criteria. A larger d-value does not ensure a good distribution of SPV values throughout the design space. It should be noted that the best design based on the FDS plots and VDGs always had the smallest v_Q value. That is, those designs are associated with the smallest estimated variances for the quadratic effects (QEs).

What is clear in these comparisons is that there exists a CBBD or a CBBD* that is superior to every BBD of the same size based on d-values, FDS plots, and VDGs. This is most likely due to the over-abundance of 0-factor levels in BBDs leading to poor prediction for larger radii.

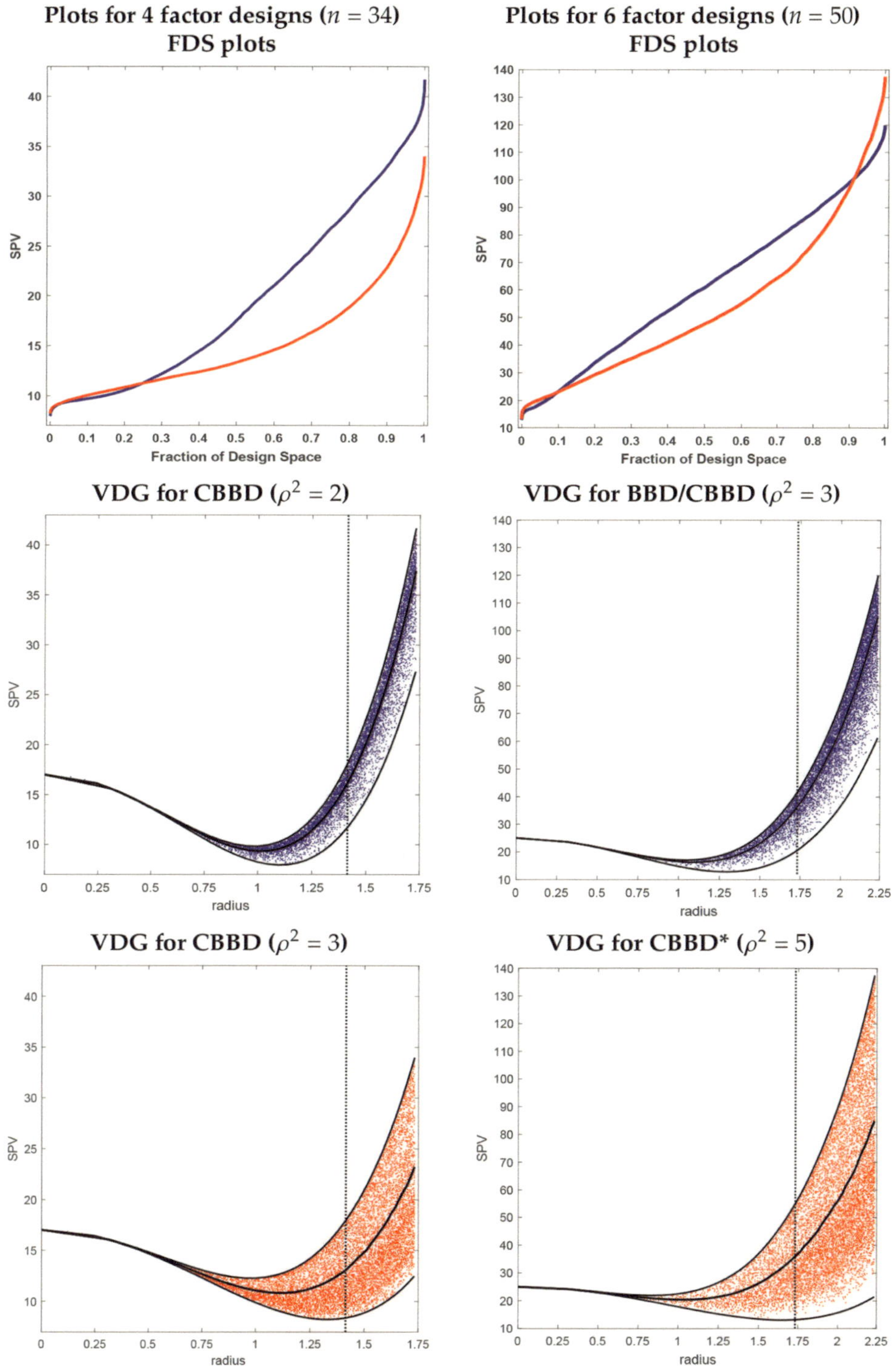

Figure 3.
FDS plots and VDGs for designs with 4 and 6 factors. FDS lines for $m = 4$: blue for $\rho^2 = 2$ CBBD and red for $\rho^2 = 3$ CBBD. FDS lines for $m = 6$: blue for $\rho^2 = 3$ BBD and red for $\rho^2 = 5$ CBBD. VDGs include solid black lines for the minimum, average, and maximum SPV. A vertical reference line is plotted at $\sqrt{2}$ for $m = 6$ and at $\sqrt{3}$ for $m = 6$.*

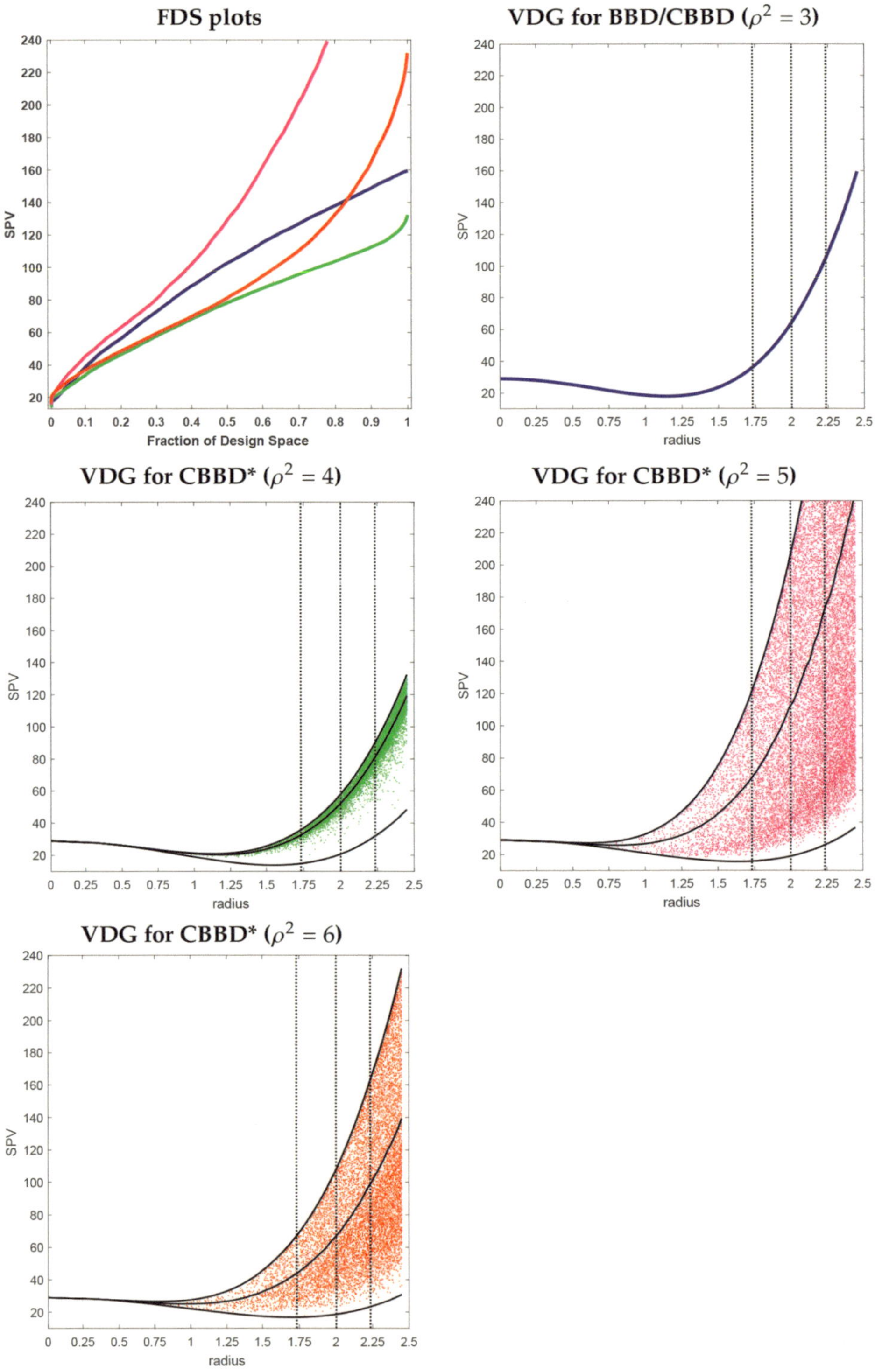

Figure 4.
*FDS plots and VDGs for designs with 7 factors ($n = 58$). FDS lines: blue for BBD/CBBD, green, magenta, and red for CBBD*s with $\rho^2 = 3,4,5,6$, respectively. VDGs include solid black lines for the minimum, average, and maximum SPV. Vertical reference lines are plotted at $\sqrt{3}$, $\sqrt{4}$ and $\sqrt{5}$.*

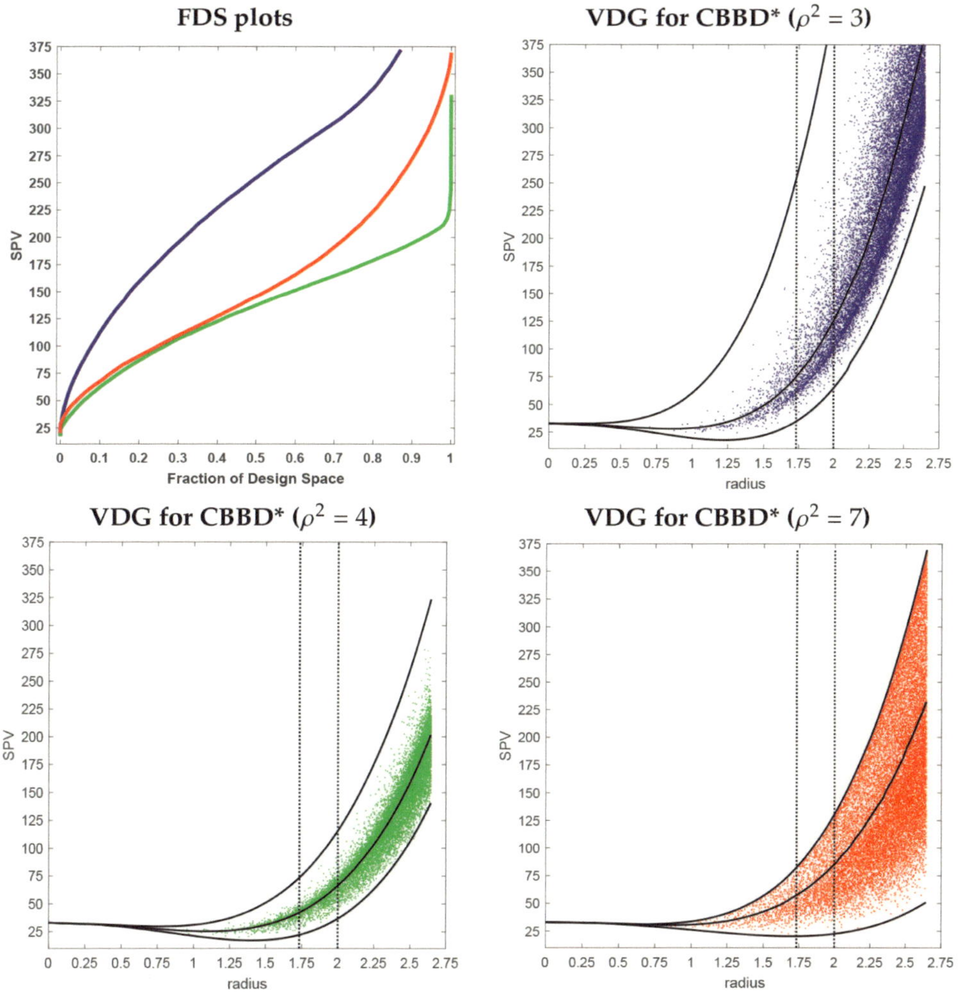

Figure 5.
*FDS plots and VDGs for designs with 8 factors ($n = 66$). FDS lines: blue, green, and red for CBBD*s with $\rho^2 = 3,4,7$, respectively. VDGs include solid black lines for the minimum, average, and maximum SPV. Vertical reference lines are plotted at $\sqrt{3}$ and $\sqrt{4}$.*

6. Conclusions

This chapter offers the cyclic-generating approach to create new designs (CBBDs and CBBD*s) as alternatives to existing BBDs. Our new designs offer a compromise between the definitive screening designs [11] (where each factor has just three 0's) and BBDs (where the number of 0's for each factor is more than the number of ± 1's). In addition to quality measures, FDS plots and VDGs were generated to assess the prediction variance properties in $(m-1)$-dimensional spherical regions. These were used to compare designs of equal size but with varying ρ^2. The comparisons indicate that for each number of design factors m, there exists a CBBD or CBBD* that is superior to a BBD based on these quality measures and graphical methods. Because of extrapolation concerns related to points extending beyond the maximum value of ρ associated with a design, it is stressed that comparisons of BBDs to CBBDs or CBBD*s should take into account for the differences in the spherical design regions based on differing ρ^2 values. Once implemented, experimental data resulting from a CBBD or CBBD* can be analyzed analogously to a data analysis for a BBD using currently available statistical software. A catalog of the RSDs in **Table 1**, which includes 15 CBBDs and nine CBBD*s is given at the link https://designcomputing.net/cbbd/.

Appendix A. Cyclic generators for the first half-fractions of 15 CBBDs in Table 1

```
# 3b

0 + +
− + 0
0 − +
− − 0

# 4b

− + 0 0
0 + 0 +
0 + 0 −
− − 0 0

# 4c

− + 0 +
+ − 0 +
− + + 0
− − − 0

# 5a

0 0 + 0 +
0 − − 0 0
+ 0 0 − 0
+ − 0 0 0

# 5b

− 0 0 − +
+ − − 0 0
0 0 + + +
− + 0 0 −

# 6a

0 + 0 0 + +
− 0 0 − + 0
0 0 + − 0 −
0 0 − − 0 +

# 7a

0 0 0 − 0 − −
+ 0 0 0 + 0 −
0 0 + 0 + − 0
0 0 0 − 0 + +

# 8b

0 − + 0 0 0 0 −
+ 0 0 0 0 + 0 −
0 + − 0 0 + 0 0
− − 0 0 − 0 0 0
0 0 0 0 + 0 + +
0 0 0 + 0 − − 0
0 + 0 0 0 − − 0
0 0 + 0 0 0 − +

# 8c

0 + 0 0 + + 0 +
0 − 0 0 − − 0 +
0 0 0 + 0 − + −
0 0 + 0 − − − 0
0 0 − + 0 + 0 +
0 0 0 − 0 − − +
0 + 0 0 − + 0 −
0 − 0 − + + 0 0

# 9b

− 0 + 0 + + 0 0 0
0 − + 0 0 0 + 0 −
0 − 0 + + 0 0 0 +
+ 0 + − 0 0 0 + 0
− − 0 0 0 − 0 − 0
− 0 0 0 + 0 + 0 −
0 + − 0 0 0 − 0 −
− 0 + − 0 0 0 + 0

# 10b

0 0 − + 0 + 0 0 + 0
+ 0 − 0 0 + 0 0 0 +
0 0 0 − − 0 − 0 0 −
0 + + 0 0 0 0 − 0 −
0 + − 0 + 0 0 − 0 0
+ 0 0 0 0 − 0 + 0 +
0 − 0 − 0 − + 0 0 0
0 − 0 + − 0 0 0 0 +

# 11b

0 − − 0 0 − 0 + 0 0 0
0 0 − 0 − 0 0 0 0 − −
0 + 0 0 0 0 − + 0 0 −
− 0 + 0 0 0 0 + + 0 0
− 0 0 − 0 + 0 0 0 0 +
0 + 0 + 0 0 0 0 − + 0
0 0 + − 0 0 + 0 + 0 0
+ 0 + 0 0 0 0 − − 0 0

# 12

0 0 + + 0 0 0 0 0 − 0 −
− − 0 0 0 0 0 + 0 − 0 0
0 0 0 − 0 + 0 0 − + 0 0
0 − 0 0 0 0 + − 0 + 0 0
0 + 0 0 + − 0 0 0 0 0 +
0 0 0 + 0 0 0 0 + + 0 +
− 0 0 0 − 0 0 0 0 + − 0
− 0 0 0 0 − − 0 + 0 0 0

# 13

0 0 0 − 0 − + 0 0 0 + 0 0
− 0 + + 0 0 0 − 0 0 0 0 0
0 0 0 − 0 0 0 0 0 − 0 − +
0 − 0 + − 0 0 0 + 0 0 0 0
− 0 0 0 0 0 + 0 − − 0 0 0
0 0 0 0 0 + 0 + + 0 0 0 +
− + 0 0 0 + 0 0 0 0 0 + 0
− − 0 0 0 + 0 0 0 0 0 − 0

# 14

0 0 0 0 0 + 0 0 0 + − 0 − 0
+ 0 0 0 0 − 0 + 0 0 0 − 0 0
+ − 0 + 0 0 0 0 0 0 − 0 0 0
0 0 0 0 − 0 0 0 − − 0 + 0 0
0 0 0 − 0 0 0 + + 0 + 0 0 0
− 0 0 0 0 − 0 − 0 0 0 − 0 0
0 + 0 0 0 + 0 0 + 0 0 0 0 −
+ 0 + 0 0 0 − 0 0 + 0 0 0 0
```

Appendix B. Cyclic generators for 9 CBBD*s in Table 1

```
# 3a

+ − 0
− 0 −
+ 0 +
+ 0 −

# 5c

+ + 0 + +
− + − 0 +
0 − − − −
− 0 − + +
− + 0 + −
− + + 0 −
+ 0 − + −
0 + + − −

# 6b

− + 0 + + +
0 − − − − +
+ + + 0 − +
− + − + − 0
+ + − 0 + −
+ − − + + 0
0 + + − − −
+ − − 0 − −

# 7b

0 + + − 0 0 −
0 0 + 0 + + +
0 0 + 0 + − −
− − + 0 0 + 0
− + 0 0 − 0 +
0 0 − 0 − + +
− 0 0 + 0 − +
− 0 − − − 0 0

# 7c

+ − 0 0 + + +
+ − + 0 0 − −
+ + − + 0 0 +
0 + − − + + 0
0 − + − − − 0
0 + − − − − 0
0 0 − − + + −
+ − + + 0 0 −

# 7d

+ + 0 − + + −
+ − 0 + + − −
− + + + + 0 +
− + − + 0 − +
0 − − − + + −
− + 0 + − − −
0 + + + + − −
− − 0 − − + −

# 8d

− − 0 + 0 0 0 0
0 + − 0 − 0 0 0
0 0 + 0 0 + − 0
0 + + 0 0 0 − 0
− 0 0 0 − 0 0 −
− + 0 − 0 0 0 0
+ 0 0 0 0 + + 0
0 + 0 0 − + 0 0

# 8e

+ 0 + + − 0 0 0
0 + 0 + 0 0 − +
0 0 − − 0 − 0 −
− 0 0 0 + 0 − −
0 0 + − 0 + 0 −
+ 0 − 0 + 0 0 +
0 − 0 − + + 0 0
0 0 − 0 + − + 0

# 8f

− + − + 0 − − +
− 0 − − − − − +
− − + + − − 0 +
+ − + + − 0 + +
0 − + − + + + +
0 − + + + − − −
− + + − + + 0 +
+ 0 + + − − − −
```

Author details

Nam-Ky Nguyen[1]*, John J. Borkowski[2] and Mai Phuong Vuong[3]

1 Vietnam Institute for Advanced Study in Mathematics, Hanoi, Vietnam

2 Montanas State University, Bozeman, MT, USA

3 Hanoi University of Science and Technology, Hanoi, Vietnam

*Address all correspondence to: nknam@viasm.edu.vn

IntechOpen

© 2022 The Author(s). Licensee IntechOpen. This chapter is distributed under the terms of the Creative Commons Attribution License (http://creativecommons.org/licenses/by/3.0), which permits unrestricted use, distribution, and reproduction in any medium, provided the original work is properly cited.

References

[1] Box GEP, Behnken DW. Some new three level second-order designs for surface fitting. Statistical Technical Research Group. 1958;**26**

[2] Box GEP, Behnken DW. Some new three level designs for the study of quantitative variables. Technometrics. 1960;**2**:455-475

[3] Box GEP, Wilson KB. On the experimental attainment of optimum conditions. Journal of the Royal Statistical Society Series B. 1951;**13**:1-45

[4] Czyrski A, Sznura J. The application of Box-Behnken design in the optimization of HPLC separation of fluoroquinolones. Scientific Reports. 2019;**9**:19458

[5] Myers RS, Montgomery DC, Anderson-Cook CM. Response surface methodology: Process and product optimization using designed experiments. In: Wiley Series in Probability and Statistics. New York: Wiley; 2016

[6] Box GEP, Draper NR. Response Surfaces, Mixtures and Ridge Analyses. 2nd ed. New York: Wiley; 2007

[7] Nguyen N-K, Pham D-T, Vuong MP. Constructing 3-level foldover screening designs using cyclic generators. Chemometrics and Intelligent Laboratory Systems. 2018;**179**:92-98

[8] Pham T-D, Nguyen N-K, Vuong MP. Constructing response surface designs with orthogonal quadratic effects using cyclic generators. Chemometrics and Intelligent Laboratory Systems. 2020; **198**:103918

[9] Nguyen N-K, Borkowski JJ. New 3-level response surface designs constructed from incomplete block designs. Journal of Statistical Planning and Inference. 2008;**138**:294-305

[10] Nguyen N-K, Vuong MP, Pham D-T. Response surface designs robust against nuisance factors. In: Kayaroganam P, editor. Response Surface Methodology in Engineering Science. London: IntechOpen; 2021

[11] Jones B, Nachtsheim CJ. A class of three levels designs for definitive screening in the presence of second order effects. Journal of Quality Technology. 2011;**43**:1-15

[12] Hussey JR, Myers RH, Houck EC. Correlated simulation experiments in first-order response surface design. Operations Research. 1987;**35**:744-758

[13] Giovannitti-Jensen A, Myers RH. Graphical assessment of the prediction capability of response surface designs. Technometrics. 1989;**31**:159-171

[14] Zahran A, Anderson-Cook CM, Myers RH. Fraction of design space to assess prediction capability of response surface designs. Journal of Quality Technology. 2003;**35**:377-386

[15] Borkowski JJ. Graphical methods for assessing the prediction capability of response surface designs. In: Response Surface Methodology and Related Topics. London: World Scientific; 2006. pp. 349-378

Section 3

Modeling and Optimization Using Response Surface Methodology

Chapter 5

Analysis and Optimization of Bead Geometry by Using Response Surface Methodology

Asif Ahmad, Shahnawaz Alam and Meenu Sharma

Abstract

Analysis of bead geometry is very important in product design and manufacturing. Defect-free products with reliability are the demanding parameter in the manufacturing Industry. In this study, we have analyzed and optimized bead geometry parameters such as height of reinforcement (HOR), depth of Heat Affected Zone (DOH), and width of Heat Affected Zone (WOH) by using Central Composite Design (CCD) of response surface methodology (RSM). In this study, peak current and pulse frequency are the most important process parameters for HOR and the optimum combination obtained are (160 A, 80 A, 100 Hz, and 45%) further HOR at this optimum was found to be 1.41 mm, which is close to 1.45 mm. Similarly, peak current and pulse frequency are the most important process parameter for WOH and the optimum combination obtained are (160 A, 80 A, 150 Hz, and 45%) further WOH at this optimum was found to be 1.32 mm, which is close to 1.37 mm. Again, similarly peak current and pulse frequency are the most important process parameter for DOH and the optimum combination obtained are (160 A, 80 A, 100 Hz, and 45%) further DOH at this optimum was found to be 1.26 mm which is close to 1.58 mm.

Keywords: bead geometry, height of reinforcement, depth of Haz, response surface methodology

1. Introduction

The traditional method of selecting one parameter is time taking process and therefore not considered nowadays in the manufacturing industry, hence an optimization technique concerning the design of experiment (DOE) such as CCD of response surface methodology (RSM) to establish an optimum condition for tensile strength. In this study, the surface plot is used to explain the main and interaction effect of the process parameter to identify the optimum parameter with their values. RSM is a widely used statistical technique in process optimization [1]. RSM is a set of mathematical and statistical methodologies for assessing problems in which multiple independent factors influence a dependent variable or response, to optimize the answer. RSM facilitates the examination of the interaction between experiment variables within the range under consideration, allowing for a better knowledge of the process while lowering experiment time and cost [2, 3].

2. Steps of Response Surface Methodology

Major steps of RSM are shown in **Figure 1**.

2.1 Input parameters and their operating range

Based on a review of the literature and previous research, the most important process parameters that have a greater influence on bead geometry and mechanical properties have been identified. The butt joint was made from AISI 316 stainless steel sheets with dimensions of 100 × 75 × 4 mm by used pulsed TIG welding [4]. This experiment's input parameters are peak current, base current, pulse frequency, and pulse on time [2]. Input parameters with their levels are given in **Table 1**. The experiment was carried out at an optimum in the laboratory.

2.2 Design of experiment

The experimental design for this investigation is CCD and the response is measured by RSM. Examine the combined effect of four different input parameters on bead geometry and mechanical properties to optimize the process parameter of pulse TIG welding and drive a mathematical model. Five levels, four-parameter CCD which include 24 = 16 factorial points plus 6 central points and 2 × 4-star points (24 + 2*4 + 6) [2, 5], with a total of 30 experiments were made in this investigation as

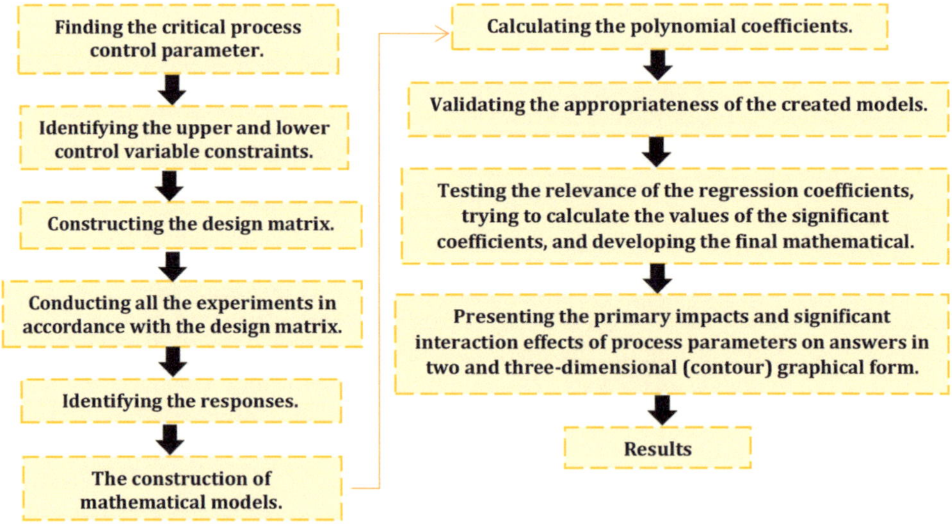

Figure 1.
Flowchart representing steps of RSM.

Input parameter	Factor symbol	Level 1	Level 2	Level 3	Level 4	Level 5
		$-\beta$	$-\gamma$	α	β	γ
Peak current (I)	A	140	150	160	170	180
Base current (I)	B	60	70	80	90	100
Pulse frequency (Hz)	C	50	75	100	125	150
Pulse on time (%)	D	35	40	45	50	55

Table 1.
Independent parameters with their levels for CCD.

shown in **Table 2**. The framework for the four factors ranged between five levels, $-\gamma$, α, $+\beta$, and $+\gamma$).

3. RSM statistical analysis for reinforcement height

By varying the input process parameter, CCD was used to experiment. The experiment was carried out by varying the input parameters with the experimental design CCD. The experiment was carried out using various parameter combinations, as shown in **Table 3**. The CCD experiment results were fitted to the polynomial regression equation created by Design Expert Software 18.0 [2, 6].

3.1 Development and evaluation of regression equation HOR

The correlation between process parameters and output response was obtained by using CCD The second-order polynomial regression equation fitted between the output response and the input process parameter. From the ANOVA result shown in **Table 4**, it has been found adequacy of the model is suitable to analyze the experimental value [2, 6].

R^2 = 0.99543, adjusted R^2 = 0.99763.

The regression equation based on the regression coefficient of ANOVA results is shown in Eq. (1).

$$\begin{aligned} \text{HOR} = {} & 1.11 + 0.0025\text{A} - 0.0275\text{B} + 0.0737\text{C} + 0.0142\text{D} + 0.0375\text{AB} \\ & - 0.0406\text{AC} + 0.0356\text{AD} + 0.0444\text{BC} - 0.0456\text{BD} + 0.0325\text{CD} \\ & - 0.0824\text{A}^2 - 0.0461\text{B}^2 - 0.0599\text{C}^2 - 0.0918\text{D}^2 \end{aligned} \tag{1}$$

To obtain a statistically significant regression model p-value, if the p-value < 0.05 then the mathematical model is significant. A, C, AB, A^2, and D^2 are significant model terms in this case. The model can be reduced to Eq. (2), after eliminating the insignificant coefficients. After that predicted value for all the combinations of input, the parameter is obtained as shown in **Table 5**.

$$\text{HOR} = 1.11 + 0.0025\text{A} + 0.0737\text{C} + 0.0375\text{AB} - 0.0824\text{A}^2 - 0.0918\text{D}^2 \tag{2}$$

Std	Factor symbol				Actual factor			
	A	B	C	D	A	B	C	D
1	$-\gamma$	$-\gamma$	$-\gamma$	$-\gamma$	150	70	75	40
2	β	$-\gamma$	$-\gamma$	$-\gamma$	170	70	75	40
3	$-\gamma$	β	$-\gamma$	$-\gamma$	150	90	75	40
4	β	β	$-\gamma$	$-\gamma$	170	90	75	40
5	$-\gamma$	$-\gamma$	β	$-\gamma$	150	70	125	40
6	β	$-\gamma$	β	$-\gamma$	170	70	125	40
7	$-\gamma$	β	β	$-\gamma$	150	90	125	40
8	β	β	β	$-\gamma$	170	90	125	40
9	$-\gamma$	$-\gamma$	$-\gamma$	β	150	70	75	50
10	β	$-\gamma$	$-\gamma$	β	170	70	75	50
11	$-\gamma$	β	$-\gamma$	β	150	90	75	50
12	β	β	$-\gamma$	β	170	90	75	50
13	$-\gamma$	$-\gamma$	β	β	150	70	125	50
14	β	$-\gamma$	β	β	170	70	125	50
15	$-\gamma$	β	β	β	150	90	125	50
16	β	β	β	β	170	90	125	50
17	$-\beta$	α	α	α	140	80	100	45
18	α	α	α	α	180	80	100	45
19	α	$-\beta$	α	α	160	60	100	45
20	α	α	α	α	160	100	100	45
21	α	α	$-\beta$	α	160	80	50	45
22	α	α	α	α	160	80	150	45
23	α	α	α	$-\beta$	160	80	100	35
24	α	α	α	α	160	80	100	55
25	α	α	α	α	160	80	100	45
26	α	α	α	α	160	80	100	45
27	α	α	α	α	160	80	100	45
28	α	α	α	α	160	80	100	45
29	α	α	α	α	160	80	100	45
30	α	α	α	α	160	80	100	45

Table 2.
Design of experiment or central composite design arrangement.

3.2 Adequacy check of the mathematical model for height of reinforcement

ANOVA represents that the polynomial regression equation was significant to represent the relationship between input parameters and output parameters. The adequacy and significance of the established model were also elaborated by the high

Std	Factor symbol				Actual factor				Exp. value Bead width
	A	B	C	D	A	B	C	D	
1	$-\gamma$	$-\gamma$	$-\gamma$	$-\gamma$	150	70	75	40	0.85
2	β	$-\gamma$	$-\gamma$	$-\gamma$	170	70	75	40	0.68
3	$-\gamma$	β	$-\gamma$	$-\gamma$	150	90	75	40	0.57
4	β	β	$-\gamma$	$-\gamma$	170	90	75	40	0.93
5	$-\gamma$	$-\gamma$	β	$-\gamma$	150	70	125	40	0.74
6	β	$-\gamma$	β	$-\gamma$	170	70	125	40	0.64
7	$-\gamma$	β	β	$-\gamma$	150	90	125	40	1.01
8	β	β	β	$-\gamma$	170	90	125	40	0.76
9	$-\gamma$	$-\gamma$	$-\gamma$	β	150	70	75	50	0.67
10	β	$-\gamma$	$-\gamma$	β	170	70	75	50	1.05
11	$-\gamma$	β	$-\gamma$	β	150	90	75	50	0.75
12	β	β	$-\gamma$	β	170	90	75	50	0.63
13	$-\gamma$	$-\gamma$	β	β	150	70	125	50	1.08
14	β	$-\gamma$	β	β	170	70	125	50	0.80
15	$-\gamma$	β	β	β	150	90	125	50	0.71
16	β	β	β	β	170	90	125	50	1.14
17	$-\beta$	α	α	α	140	80	100	45	0.88
18	α	α	α	α	180	80	100	45	0.78
19	α	$-\beta$	α	α	160	60	100	45	1.14
20	α	α	α	α	160	100	100	45	0.81
21	α	α	$-\beta$	α	160	80	50	45	0.67
22	α	α	α	α	160	80	150	45	1.17
23	α	α	α	$-\beta$	160	80	100	35	0.87
24	α	α	α	α	160	80	100	55	0.71
25	α	α	α	α	160	80	100	45	1.31
26	α	α	α	α	160	80	100	45	0.92
27	α	α	α	α	160	80	100	45	0.76
28	α	α	α	α	160	80	100	45	1.07
29	α	α	α	α	160	80	100	45	1.18
30	α	α	α	α	160	80	100	45	1.45

Table 3.
CCD experimental value: HOR.

value of the coefficient of determination (R^2) value of 0.99543 and adjusted R^2 0.99763 for the development of the developed correlation [2, 3].

Figure 2 demonstrates that the regression model generated with Design Expert 18.0 has a good correlation between the experimental and predicted values since all of the points are very close to the line of perfect fit or line of unit slope. Furthermore, residuals were investigated to validate the model's adequacy. The difference between

Source	Coefficient	Sum of squares	df	Mean square	*F*-values	*p*-value	
Model	1.11	0.7246	14	0.7246	1.10	0.0425	Significant
A	0.0024	0.0003	1	0.0003	3.003	0.047	
B	−0.0275	0.0181	1	0.0181	0.087	0.5331	
C	0.0691	0.1307	1	0.1307	2.79	0.0115	
D	0.0144	0.0044	1	0.0044	0.1028	0.7456	
A × B	0.0375	0.0235	1	0.0235	0.4609	0.0498	
A × C	−0.0406	0.0276	1	0.0276	0.5849	0.4643	
A × D	0.0356	0.0208	1	0.0208	0.4268	0.5107	
B × C	0.0444	0.0328	1	0.0328	0.5729	0.4359	
B × D	−0.0456	0.0356	1	0.0356	0.7023	0.4123	
C × D	0.0325	0.0168	1	0.0168	0.3508	0.5670	
A^2	−0.0824	0.1862	1	0.1862	3.98	0.0447	
B^2	−0.0461	0.0584	1	0.0584	1.25	0.2817	
C^2	−0.0599	0.0984	1	0.0984	2.10	0.1678	
D^2	−0.0918	0.2310	1	0.2310	4.93	0.0422	
Residual		0.7025	15	0.0468			
Lack of fit		0.3837	10	0.0384	0.6017	0.7691	Not significant
Pure error		0.3188	5	0.0638			
Cor total	1.11	0.7230	14	0.0516	1.10	0.0425	Significant

Table 4.
ANOVA for the: HOR.

the observed and predicted responses is referred to as the residual. This analysis was examined using the normal probability plot of residuals [2, 5]. The normal probability plot of the residuals shows that the errors are distributed normally in a straight line and are insignificant as shown in **Figure 3**.

3.3 Perturbation plot: height of reinforcement

The perturbation plot shows the effect of all the parameters on a single plot A perturbation plot to compare the effect of all the process parameters at the center point on bead width is presented in **Figure 4**. It has been noted that HOR peak current (A) is increasing and then HOR decreases [1, 6].

The plot also shows that the HOR decreases as the base current (B) increases because no melting occurs during this stage. This plot shows that HOR increases as pulse frequency (C) increases. The plot also shows that HOR increases as pulse on-time increases (D) and then decreases [2].

3.4 Response surface plot: height of reinforcement

The 3D surface plot and 2D contour effect developed by design expert 18.0 software represent the interaction effect between process parameters and HOR as shown in **Figures 5–10** [3].

Std	Factor sign				Estimated value	Remaining error
	A	B	C	D		
1	$-\gamma$	$-\gamma$	$-\gamma$	$-\gamma$	0.849	0.012
2	β	$-\gamma$	$-\gamma$	$-\gamma$	0.789	−0.098
3	$-\gamma$	β	$-\gamma$	$-\gamma$	0.722	−0.136
4	β	β	$-\gamma$	$-\gamma$	0.812	0.134
5	$-\gamma$	$-\gamma$	β	$-\gamma$	0.924	−0.168
6	β	$-\gamma$	β	$-\gamma$	0.702	−0.046
7	$-\gamma$	β	β	$-\gamma$	0.974	0.047
8	β	β	β	$-\gamma$	0.902	−0.126
9	$-\gamma$	$-\gamma$	$-\gamma$	β	0.832	−0.151
10	β	$-\gamma$	$-\gamma$	β	0.915	0.146
11	$-\gamma$	β	$-\gamma$	β	0.522	0.239
12	β	β	$-\gamma$	β	0.755	−0.109
13	$-\gamma$	$-\gamma$	β	β	1.037	0.059
14	β	$-\gamma$	β	β	0.957	−0.141
15	$-\gamma$	β	β	β	0.905	−0.179
16	β	β	β	β	0.975	0.181
17	$-\beta$	α	α	α	0.794	0.097
18	α	α	α	α	0.804	−0.013
19	α	$-\beta$	α	α	0.999	0.152
20	α	α	α	α	0.889	−0.068
21	α	α	$-\beta$	α	0.741	−0.060
22	α	α	α	α	1.036	0.145
23	α	α	α	$-\beta$	0.733	0.148
24	α	α	α	α	0.790	−0.064
25	α	α	α	α	1.129	0.193
26	α	α	α	α	1.129	−0.193
27	α	α	α	α	1.129	−0.358
28	α	α	α	α	1.129	−0.043
29	α	α	α	α	1.129	0.068
30	α	α	α	α	1.129	0.333

Table 5.
CCD predicted value: HOR.

The coefficient of the linear interactive effect of peak current and base current is positive as given in **Table 4**. HOR is increased as the value of the above parameter is increased as shown in **Figure 5a** of the 3D surface plot and **Figure 5b** of the contour plot. HOR increases with concurrent increases in peak current and base current to

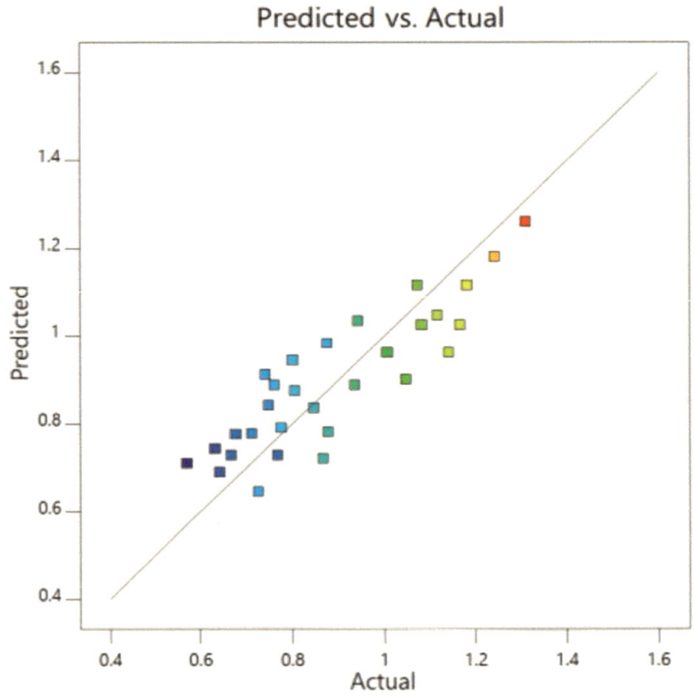

Figure 2.
Plot of experimental vs. predicted value HOR.

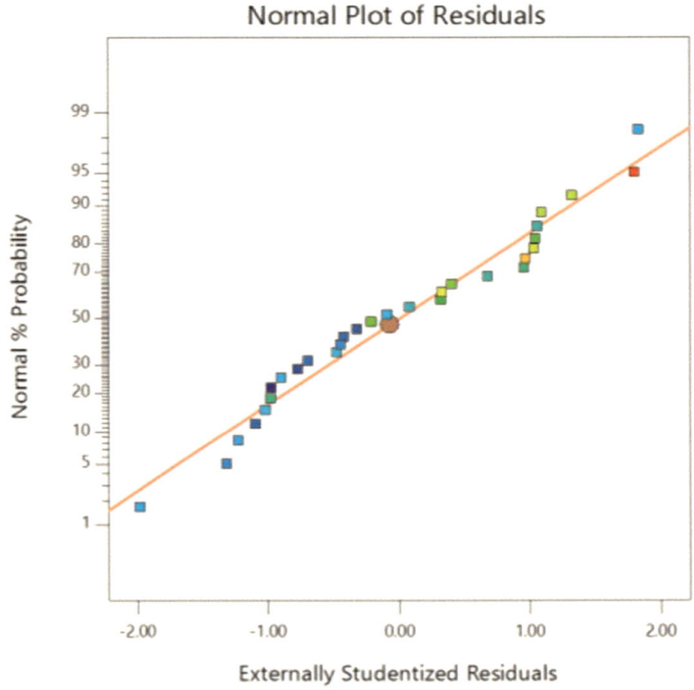

Figure 3.
Normal probability plot of residual HOR.

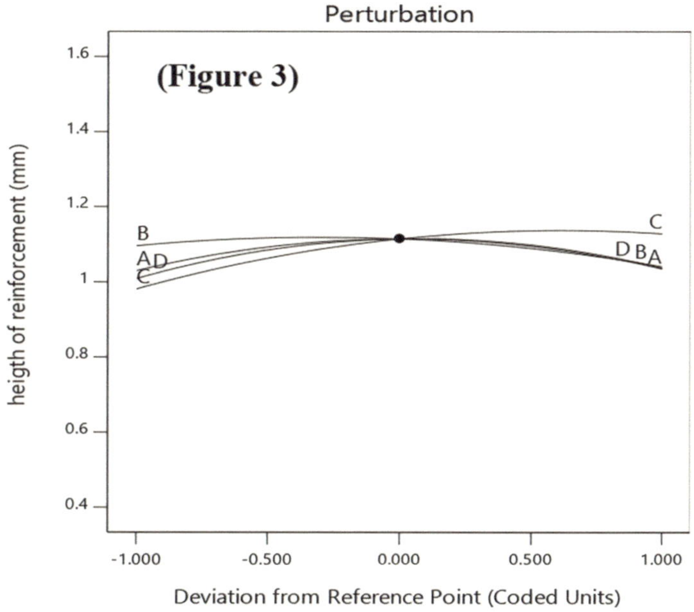

Figure 4.
Perturbation plot of HOR.

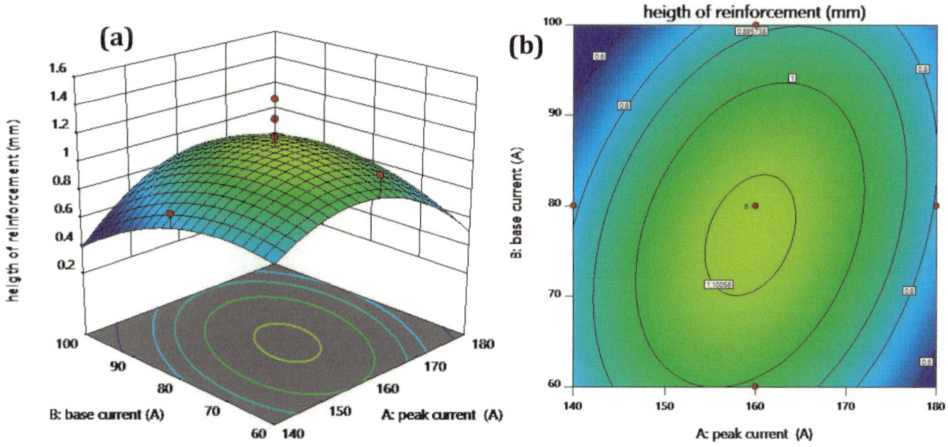

Figure 5.
Surface plot (a), contour plot (b) of the interaction effect AB on HOR.

approximately 180–100 A, respectively, beyond which the value of HOR decreases [2, 3]. As shown in **Table 4**, the coefficient of linear interactive effects of peak current and pulse frequency is negative. As shown in **Figure 6c** of 3D surface plots and **Figure 6d** of contour plots, HOR increases as the value of the above parameter increases. The HOR declined beyond the peak current of 180 A and pulse frequency of 125 Hz respectively [2, 6].

As shown in **Table 4**, the coefficient of the linear effect of peak current and pulse on time is positive. As shown in **Figure 7e** of the 3D surface plot and **Figure 7f** of the contour plot, HOR increases as the value of the above parameter increases. DOP is

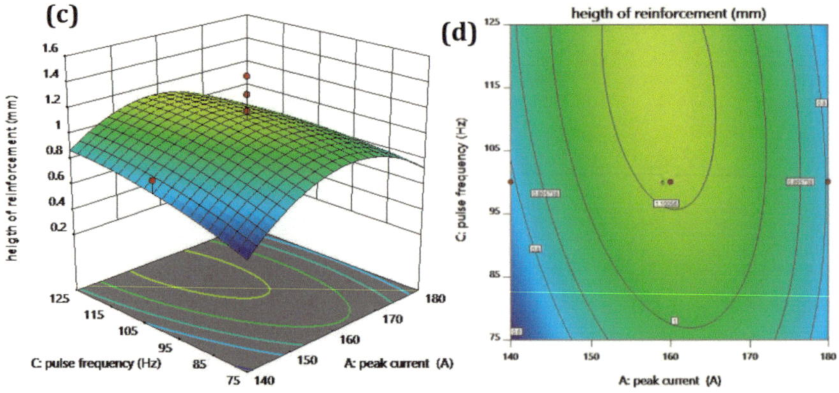

Figure 6.
Surface plot (c), contour plot (d) of the interaction effect AC on HOR.

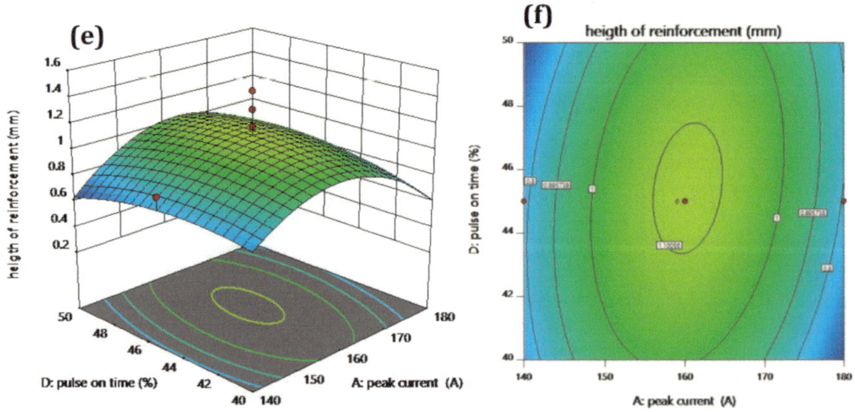

Figure 7.
Surface plot (e), contour plot (f) of the interaction effect AD ion HOR.

increasing with simultaneously increasing in peak current and pulse on time to about 180 A and 50% respectively beyond which the value of HOR declined. **Table 4** shows that the coefficient of the linear effect of base current and pulse frequency is positive. As shown in **Figure 8g** of the 3D surface plot and **Figure 8h** of the contour plot, HOR increases as the value of the above parameter increases. DOP rises as peak current and pulse on time rise to around 180 A and 50%, respectively, beyond which the value of HOR tends to fall. **Table 4** shows that the coefficient of linear interactive effects of base current and pulse on time is negative [2]. As shown in **Figure 9i** of the 3D surface plots and **Figure 9j** of the contour plot, BW increases as the value of the above parameter increases. Beyond the base current of 100 A, the HOR and pulse on time both decreased by 50%. The coefficient of the linear interactive effect of pulse frequency and pulse on time is positive as given in **Table 4**. As the value of the above parameter is increased, BW increases, as shown in **Figure 10k** of the 3D surface plot and **Figure 10l** of the contour plot [1, 2]. HOR is increasing with simultaneously increasing pulse frequency and pulse on time to about 125 Hz and 50% respectively beyond which the value of HOR declined.

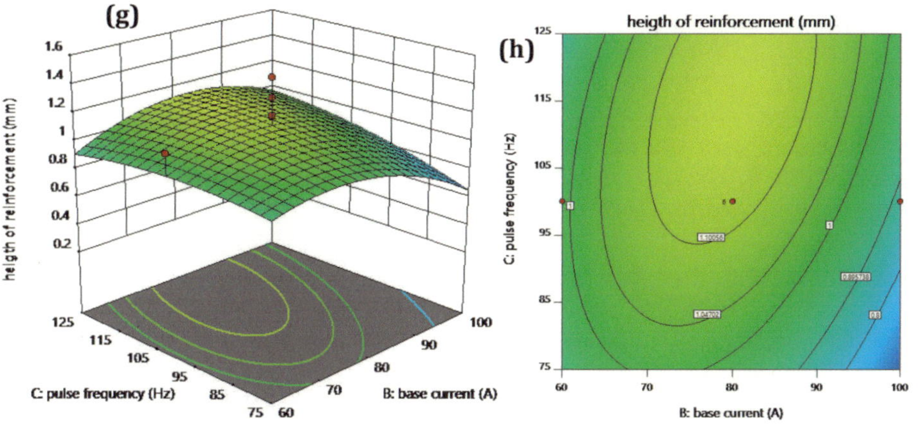

Figure 8.
Surface plot (g), contour plot (h) of the interaction effect BC ion HOR.

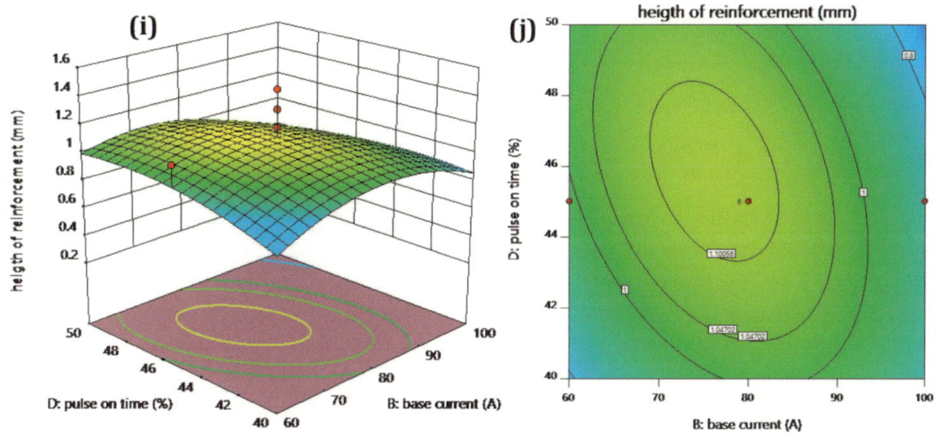

Figure 9.
Surface plot (i), contour plot (j) of the interaction effect BD on HOR.

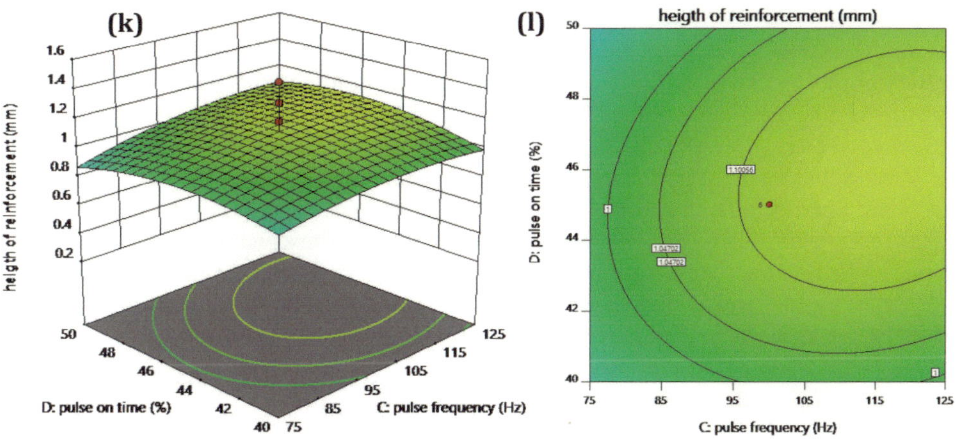

Figure 10.
Surface plot (k), contour plot (l) of the interaction effect CD on HOR.

4. Statistical analysis for depth of heat affected zone using RSM

By varying the input process parameter, CCD was used to experiment. The experiment was carried out by varying the input parameters with the experimental design CCD. The experiment was carried out using various parameter combinations, as shown in **Table 6**. The CCD experiment results were fitted to the polynomial regression equation created by Design Expert Software 18.0 [2, 5].

4.1 Development and evaluation of regression equation: depth of HAZ

The correlation between process parameters and output response was obtained by using CCD. The second-order polynomial regression equation fitted between the output response and input process parameter. From the ANOVA result shown in **Table 7**, it has been found adequacy of the model is suitable to analyze the experimental value.

R^2 = 0.98346, adjusted R^2 = 0.98459.

The regression equation based on the regression coefficient of ANOVA results is shown in Eq. (3).

$$\begin{aligned} \text{DOH} = {} & 1.12 + 0.0087\text{A} - 0.0529\text{B} + 0.1225\text{C} + 0.0400\text{D} + 0.0594\text{AB} \\ & - 0.0562\text{AC} + 0.0563\text{AD} + 0.0594\text{BC} - 0.0769\text{BD} + 0.0538\text{CD} \\ & - 0.0976\text{A}^2 - 0.0095\text{B}^2 - 0.0120\text{C}^2 - 0.0920\text{D}^2 \end{aligned} \quad (3)$$

To obtain a statistically significant regression model *p*-value, if the *p*-value < 0.05 then the mathematical model is significant. In this case, A, C, and BC^2 are significant model terms. The model reduces to Eq. (4), after eliminating the insignificant coefficients. After that predicted value for all the combinations of input parameters is obtained as shown in **Table 8**.

$$\text{DOH} = 1.12 + 0.0087\text{A} + 0.1225\text{C} + 0.0594\text{BC}^2 \quad (4)$$

4.2 Adequacy check of the mathematical model for depth of HAZ

ANOVA represents that the polynomial regression equation was significant to represent the relationship between input parameters and output parameters. The adequacy and significance of the established model were also elaborated by the high value of the coefficient of determination (R^2) value of 0.98346 and adjusted R^2 0.98459 for the development of the developed correlation. **Figure 11** shows that the regression model created with Design Expert 18.0 has a good correlation between the experimental and predicted values because all of the points are very close to the line of perfect fit or line of unit slope. Furthermore, residuals were investigated to validate the model's adequacy. The difference between the observed and predicted responses is referred to as the residual. The normal probability plot of residuals was used to examine this analysis [2, 3]. The normal probability plot of the residuals shows that the errors are distributed normally in a straight line and are insignificant as shown in **Figure 12**.

4.3 Perturbation plot: depth of heat affected zone

The perturbation plot shows the effect of all the parameters on a single plot. **Figure 13** shows a perturbation plot that compares the effect of all process parameters

Std	Factor symbol				Experimental value Bead width
	A	B	C	D	
1	$-\gamma$	$-\gamma$	$-\gamma$	$-\gamma$	0.87
2	β	$-\gamma$	$-\gamma$	$-\gamma$	0.60
3	$-\gamma$	β	$-\gamma$	$-\gamma$	0.52
4	β	β	$-\gamma$	$-\gamma$	0.98
5	$-\gamma$	$-\gamma$	β	$-\gamma$	0.67
6	β	$-\gamma$	β	$-\gamma$	0.59
7	$-\gamma$	β	β	$-\gamma$	1.07
8	β	β	β	$-\gamma$	0.76
9	$-\gamma$	$-\gamma$	$-\gamma$	β	0.62
10	β	$-\gamma$	$-\gamma$	β	1.25
11	$-\gamma$	β	$-\gamma$	β	0.74
12	β	β	$-\gamma$	β	0.62
13	$-\gamma$	$-\gamma$	β	β	1.33
14	β	$-\gamma$	β	β	0.83
15	$-\gamma$	β	β	β	0.68
16	β	β	β	β	1.37
17	$-\beta$	α	α	α	0.94
18	α	α	α	α	0.80
19	α	$-\beta$	α	α	1.53
20	α	α	α	α	0.92
21	α	α	$-\beta$	α	0.74
22	α	α	α	α	1.68
23	α	α	α	$-\beta$	0.97
24	α	α	α	α	0.76
25	α	α	α	α	1.65
26	α	α	α	α	1.05
27	α	α	α	α	0.78
28	α	α	α	α	1.00
29	α	α	α	α	1.05
30	α	α	α	α	1.16

Table 6.
CCD experimental value: depth of HAZ (DoH).

at the center point on bead width. It has been observed that HOR peak current (A) increases before decreasing. The plot also shows that the HOR decreases as the base current (B) increases because no melting occurs during this stage. This plot shows that HOR increases as the pulse frequency (C) increases. The plot also shows that HOR increases as a pulse on time increases (D) and then decreases [2].

Source	Coefficient	Sum of squares	df	Mean square	*F*-values	*p*-value	
Model	1.12	1.26	14	0.0902	0.8929	0.5814	Significant
A	0.0087	0.0018	1	0.0018	4.71	0.045	
B	−0.0529	0.0672	1	0.0672	0.3801	0.4275	
C	0.1225	0.3601	1	0.3601	3.56	0.0478	
D	0.0400	0.0384	1	0.0384	0.6652	0.5468	
A × B	0.0594	0.0564	1	0.0564	0.5583	0.4665	
A × C	−0.0562	0.0506	1	0.0506	0.5011	0.4899	
A × D	0.0563	0.0506	1	0.0506	0.5011	0.4899	
B × C	0.0594	0.0564	1	0.0564	0.5583	0.0466	
B × D	−0.0769	0.0946	1	0.0946	0.9359	0.3487	
C × D	0.0538	0.0462	1	0.0462	0.4575	0.5091	
A^2	−0.0976	0.2613	1	0.2613	2.59	0.1286	
B^2	−0.0095	0.0025	1	0.0025	0.0244	0.8780	
C^2	−0.0120	0.0039	1	0.0039	0.0390	0.8462	
D^2	−0.0920	0.2321	1	0.2321	2.30	0.1504	
Residual		1.52	15	0.1010			
Lack of fit		1.19	10	0.1189	1.82	0.2639	Not significant
Pure error		0.3266	5	0.0653			
Cor total		2.78	29				

Table 7.
ANOVA: depth of HAZ.

4.4 Response surface plot: depth of heat affected zone

The 3D surface plot and 2D contour effect developed by design expert 18.0 software represent the interaction effect between process parameters and BW as shown in **Figures 14–19**.

The coefficient of the linear interactive effect of peak current and base current is +ve as given in **Table 7**, DOH is increased as the value of the above parameter is increased as shown in **Figure 14a** of the 3D surface plot and **Figure 14b** of the contour plot. DOH rises in tandem with increases in peak and base current to around 180 and 100 A, respectively, after which the value of DOH falls. **Table 7** shows that the coefficients of linear effects of peak current and pulse frequency are negative. As shown in **Figure 15c** of 3D surface plots and **Figure 15d** of contour plots, DOH increases as the value of an above parameter increases. The DOH decreased after reaching a peak current of 180 A and a pulse frequency of 125 Hz. The linear effect of peak current and pulse on time has a positive coefficient, as shown in **Table 7**, and DOH increases as the value of the above parameter increases, as shown in **Figure 16e** of the 3D surface plot and **Figure 16f** of the contour plot [1, 2]. DOH is increasing with simultaneously increasing in peak current and pulse on time to about 180 A and 50% respectively beyond which the value of DOH declines.

The coefficient of the linear interactive effect of base current and pulse frequency is positive as given in **Table 7**. DOH is increased as the value of the above parameter is

Std	Factor sign				Estimated value	Remaining error
	A	B	C	D		
1	$-\gamma$	$-\gamma$	$-\gamma$	$-\gamma$	0.89	−0.01625
2	β	$-\gamma$	$-\gamma$	$-\gamma$	0.79	−0.185
3	$-\gamma$	β	$-\gamma$	$-\gamma$	0.70	−0.18167
4	β	β	$-\gamma$	$-\gamma$	0.84	0.142083
5	$-\gamma$	$-\gamma$	β	$-\gamma$	1.02	−0.3475
6	β	$-\gamma$	β	$-\gamma$	0.69	−0.10625
7	$-\gamma$	β	β	$-\gamma$	1.07	0.004583
8	β	β	β	$-\gamma$	0.98	−0.21667
9	$-\gamma$	$-\gamma$	$-\gamma$	β	0.90	−0.28
10	β	$-\gamma$	$-\gamma$	β	1.03	0.22125
11	$-\gamma$	β	$-\gamma$	β	0.41	0.332083
12	β	β	$-\gamma$	β	0.77	−0.14917
13	$-\gamma$	$-\gamma$	β	β	1.25	0.08375
14	β	$-\gamma$	β	β	1.15	−0.315
15	$-\gamma$	β	β	β	0.99	−0.31167
16	β	β	β	β	1.13	0.242083
17	$-\beta$	α	α	α	0.71	0.222917
18	α	α	α	α	0.75	0.047917
19	α	$-\beta$	α	α	1.19	0.337083
20	α	α	α	α	0.98	−0.06625
21	α	α	$-\beta$	α	0.83	−0.07708
22	α	α	α	α	1.32	0.347917
23	α	α	α	$-\beta$	0.67	0.317917
24	α	α	α	α	0.83	−0.04708
25	α	α	α	α	1.12	0.49
26	α	α	α	α	1.12	−0.04
27	α	α	α	α	1.12	−0.245
28	α	α	α	α	1.12	−0.125
29	α	α	α	α	1.12	−0.095
30	α	α	α	α	1.12	0.015

Table 8.
CCD: predicted value.

increased as shown in **Figure 17g** of the 3D surface plot and **Figure 17h** of the contour plot. DOH rises as the base current and pulse frequency rise to around 100 A and 125 Hz, respectively, beyond which the value of DOH falls. As shown in **Table 7**, the coefficient of the linear effect of base current and pulse frequency is positive. As shown in **Figure 18i** of the 3D surface plot and **Figure 18j** of the contour plot, DOH

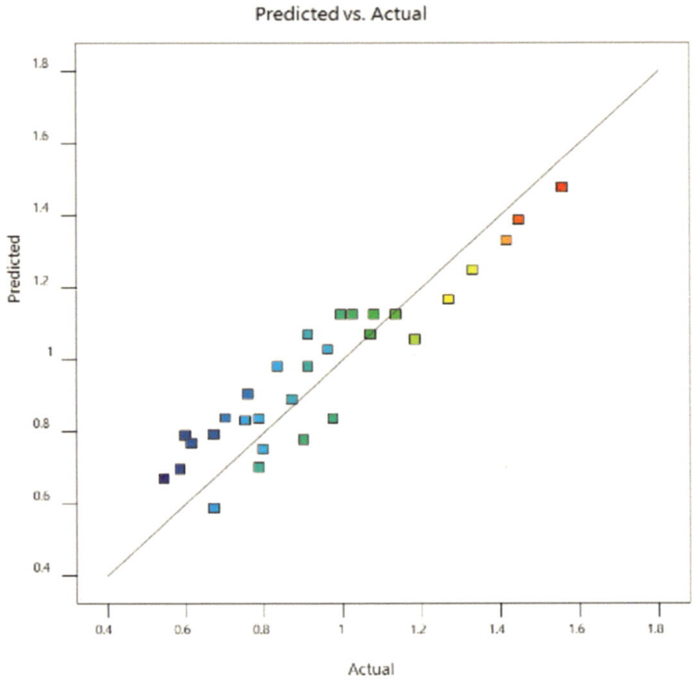

Figure 11.
Plot of experimental vs. predicted value DOH.

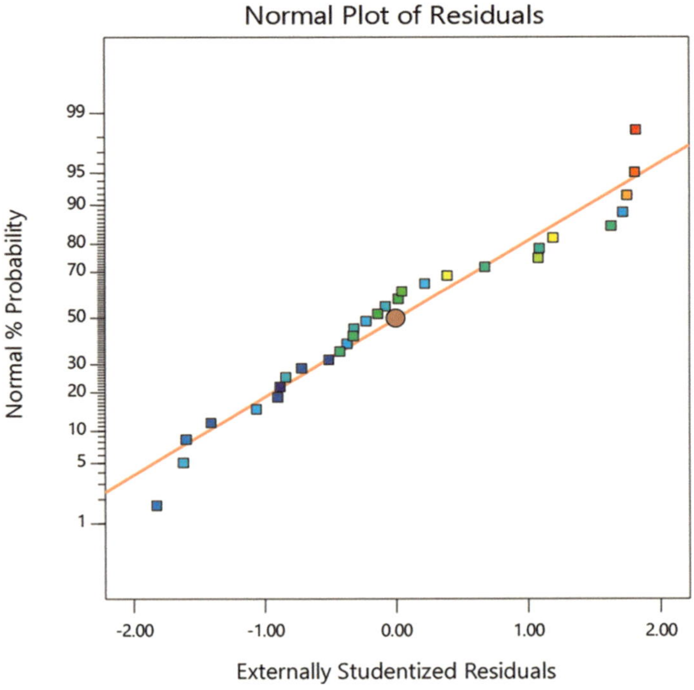

Figure 12.
Normal probability plot of residual DOH.

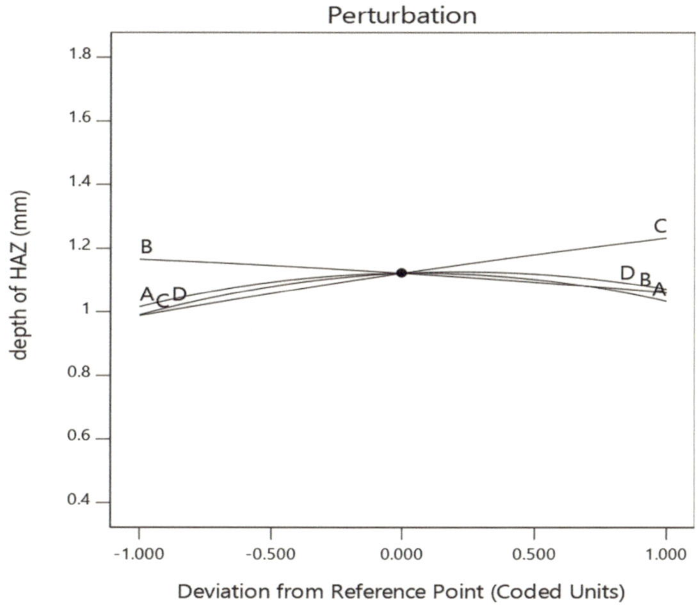

Figure 13.
Perturbation plot of DOH.

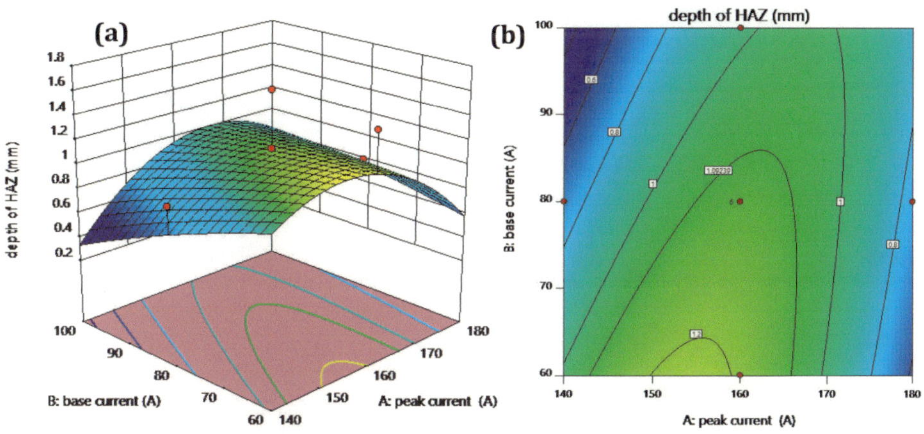

Figure 14.
Surface plot (a), contour plot (b) of the interaction effect AB on DOH.

increases as the value of the above parameter increases [1, 2]. DOH is increasing with simultaneously increasing base current and pulse frequency to about 100 A and 125 Hz respectively beyond which the value of DOH decline.

Table 7 shows that the coefficient of linear effects of base current and pulse on time is −ve. As shown in **Figure 19k** of 3D surface plots and **Figure 19l** of contour plots, DOH increases as the value of the above parameter increases. The DOH declined beyond the base current of 100 A and pulse on time by 50% respectively [2, 3].

Table 7 shows that the coefficient of the linear effect of pulse frequency and pulse on time is positive. As shown in **Figure 19k** of the 3D surface plot and **Figure 19l** of the contour plot, DOH increases as the value of the above parameter increases [2, 6].

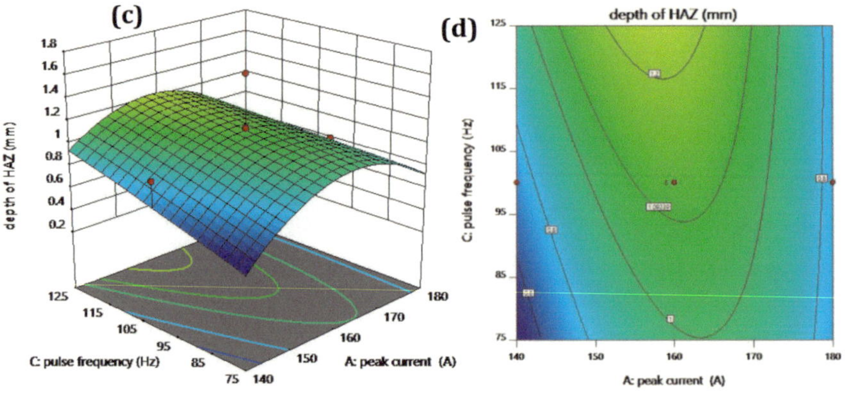

Figure 15.
Surface plot (c), contour plot (d) of the interaction effect AC on DOH.

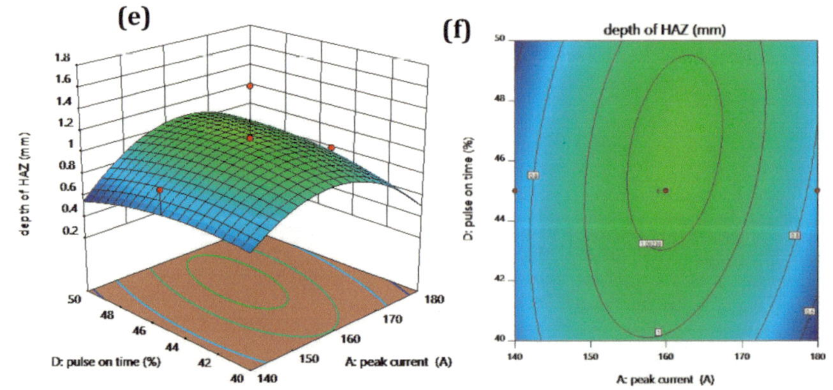

Figure 16.
Surface plot (e), contour plot (f) of the interaction effect AD on DOH.

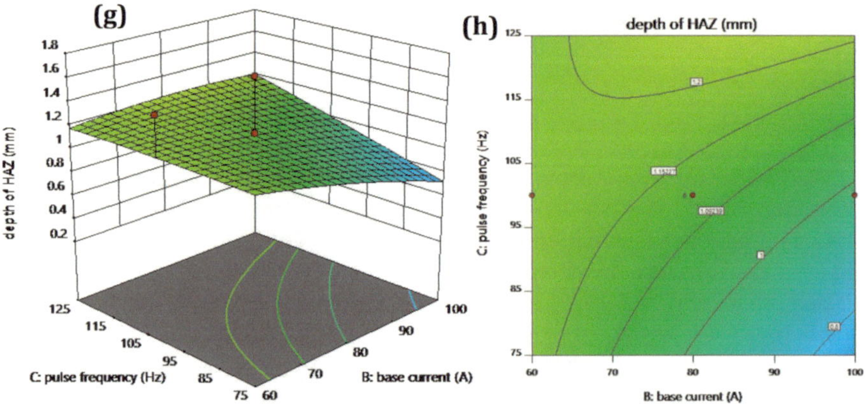

Figure 17.
Surface plot (g), contour plot (h) of the interaction effect BC on DOH.

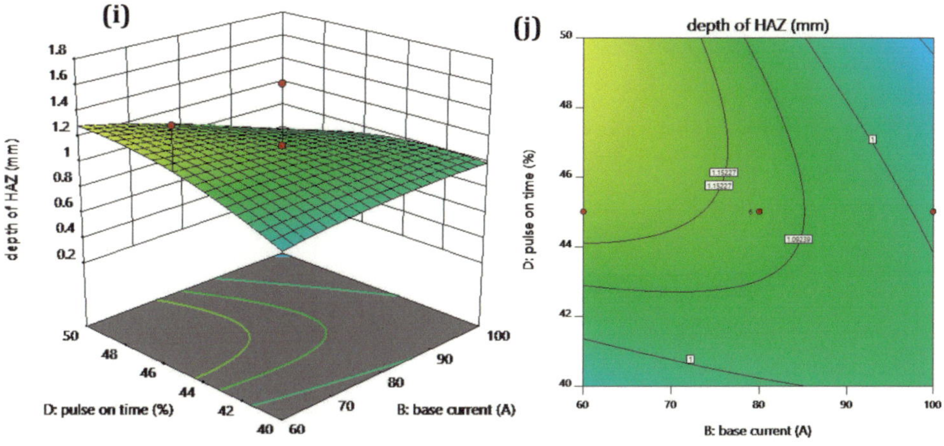

Figure 18.
Surface plot (i), contour plot (j) of the interaction effect BD on DOH.

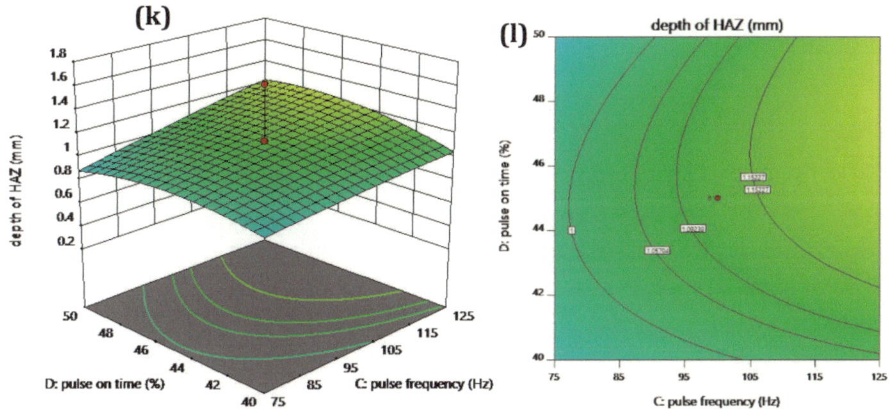

Figure 19.
Surface plot (k), contour plot (l) of the interaction effect CD on DOH.

DOH is increasing with simultaneously increasing pulse frequency and pulse on time to about 100 Hz and 50% respectively beyond which the value of DOH declines.

5. Statistical analysis for the width of heat affected zone using RSM

CCD was used to experiment by changing the input process parameter. The experiment was carried out by varying the input parameters using the experimental design CCD. The experiment was carried out using various parameter combinations, as shown in **Table 9**. The CCD experiment results were fitted to the polynomial regression equation created by Design Expert Software 18.0 [1, 2].

5.1 Development and evaluation of regression equation: width of HAZ

The correlation between process parameters and output response was obtained by using CCD. The second-order polynomial regression equation fitted between the

Std	Factor symbol				Exp. value WOH
	A	B	C	D	
1	$-\gamma$	$-\gamma$	$-\gamma$	$-\gamma$	0.92
2	β	$-\gamma$	$-\gamma$	$-\gamma$	0.68
3	$-\gamma$	β	$-\gamma$	$-\gamma$	0.57
4	β	β	$-\gamma$	$-\gamma$	1.11
5	$-\gamma$	$-\gamma$	β	$-\gamma$	0.82
6	β	$-\gamma$	β	$-\gamma$	0.67
7	$-\gamma$	β	β	$-\gamma$	1.23
8	β	β	β	$-\gamma$	0.93
9	$-\gamma$	$-\gamma$	$-\gamma$	β	0.74
10	β	$-\gamma$	$-\gamma$	β	1.13
11	$-\gamma$	β	$-\gamma$	β	0.87
12	β	β	$-\gamma$	β	0.70
13	$-\gamma$	$-\gamma$	β	β	1.21
14	β	$-\gamma$	β	β	0.95
15	$-\gamma$	β	β	β	0.79
16	β	β	β	β	1.40
17	$-\beta$	α	α	α	1.02
18	α	α	α	α	0.90
19	α	$-\beta$	α	α	1.36
20	α	α	α	α	0.98
21	α	α	$-\beta$	α	0.77
22	α	α	α	α	1.45
23	α	α	α	$-\beta$	1.08
24	α	α	α	α	0.88
25	α	α	α	α	1.28
26	α	α	α	α	1.36
27	α	α	α	α	1.09
28	α	α	α	α	1.15
29	α	α	α	α	1.19
30	α	α	α	α	1.21

Table 9.
CCD experimental value: width of HAZ (WoH).

output response and input process parameter. From the ANOVA result shown in **Table 10**, it has been found adequacy of the model is suitable to analyze the experimental value [2, 3].

R^2 = 0.9697, adjusted R^2 = 0.9734.

The regression equation based on the regression coefficient of ANOVA results is shown in Eq. (5).

Source	Coefficient	Sum of squares	df	Mean square	*F*-values	*p*-value	
Model	1.26	1.04	14	0.0741	1.26	0.3307	Significant
A	0.0079	0.0015	1	0.0015	5.025	0.0463	
B	−0.0125	0.0038	1	0.0038	0.0163	0.8042	
C	0.1108	0.2948	1	0.2948	5.01	0.0408	
D	0.0192	0.0088	1	0.0088	0.6498	0.7042	
A × B	0.0581	0.0541	1	0.0541	0.9183	0.0353	
A × B	−0.0394	0.0248	1	0.0248	0.4214	0.5261	
A × D	0.0463	0.0342	1	0.0342	0.5814	0.4576	
B × C	0.0575	0.0529	1	0.0529	0.8986	0.3582	
B × D	−0.0631	0.0638	1	0.0638	1.08	0.3145	
C × D	0.0344	0.0189	1	0.0189	0.3212	0.5793	
A^2	−0.0978	0.2624	1	0.2624	4.46	0.0419	
B^2	−0.0453	0.0563	1	0.0563	0.9567	0.3435	
C^2	−0.0603	0.0998	1	0.0998	1.69	0.2126	
D^2	−0.0922	0.2331	1	0.2331	3.96	0.0651	
Residual		0.8830	15	0.0589			
Lack of fit		0.7191	10	0.0719	2.19	0.1997	Not significant
Pure error		0.1639	5	0.0328			
Cor total		1.92	29				

Table 10.
ANOVA: WoH.

$$\begin{aligned} \text{WOH} = {} & 1.26 + 0.0079\text{A} - 0.0125\text{B} + 0.1108\text{C} + 0.0192\text{D} + 0.0581\text{AB} \\ & - 0.0394\text{AC} + 0.0463\text{AD} + 0.0575\text{BC} - 0.0631\text{BD} + 0.0344\text{CD} \\ & - 0.0978\text{A}^2 - 0.0453\text{B}^2 - 0.0603\text{C}^2 - 0.0922\text{D}^2 \end{aligned} \tag{5}$$

To obtain a statistically significant regression model p-value, if the p-value < 0.05 then the mathematical model is significant. In this case, A, C, AB, and A^2 are significant model terms. The model reduces to Eq. (6), after eliminating the insignificant coefficients. After that predicted value for all the combinations of input parameters is obtained as shown in **Table 11**.

$$\text{WOH} = 1.26 + 0.0079\text{A} + 0.1108\text{C} + 0.0581\text{AB} - 0.0978\text{A}^2 \tag{6}$$

5.2 Adequacy check of the mathematical model for the width of HAZ

ANOVA represents that the polynomial regression equation was significant to represent the relationship between input parameters and output parameters. The high value of the coefficient of determination (R^2) value of 0.9697 and the adjusted R^2 of 0.9734 for the development of the developed correlation further elaborated the adequacy and significance of the established model. **Figure 20** shows that the regression

Std	Factor sign				Estimated value	Remaining error
	A	B	C	D		
1	$-\gamma$	$-\gamma$	$-\gamma$	$-\gamma$	0.93	−0.01
2	β	$-\gamma$	$-\gamma$	$-\gamma$	0.82	−0.14
3	$-\gamma$	β	$-\gamma$	$-\gamma$	0.80	−0.23
4	β	β	$-\gamma$	$-\gamma$	0.92	0.18
5	$-\gamma$	$-\gamma$	β	$-\gamma$	1.05	−0.23
6	β	$-\gamma$	β	$-\gamma$	0.78	−0.11
7	$-\gamma$	β	β	$-\gamma$	1.15	0.08
8	β	β	β	$-\gamma$	1.11	−0.18
9	$-\gamma$	$-\gamma$	$-\gamma$	β	0.94	−0.20
10	β	$-\gamma$	$-\gamma$	β	1.01	0.12
11	$-\gamma$	β	$-\gamma$	β	0.55	0.31
12	β	β	$-\gamma$	β	0.86	−0.16
13	$-\gamma$	$-\gamma$	β	β	1.19	0.02
14	β	$-\gamma$	β	β	1.10	−0.15
15	$-\gamma$	β	β	β	1.04	−0.25
16	β	β	β	β	1.18	0.22
17	$-\beta$	α	α	α	0.85	0.16
18	α	α	α	α	0.88	0.02
19	α	$-\beta$	α	α	1.10	0.26
20	α	α	α	α	1.05	−0.08
21	α	α	$-\beta$	α	0.80	−0.03
22	α	α	α	α	1.24	0.21
23	α	α	α	$-\beta$	0.85	0.23
24	α	α	α	α	0.93	−0.05
25	α	α	α	α	1.26	0.32
26	α	α	α	α	1.26	0.10
27	α	α	α	α	1.26	−0.17
28	α	α	α	α	1.26	−0.11
29	α	α	α	α	1.26	−0.07
30	α	α	α	α	1.26	−0.05

Table 11.
CCD: predicted value.

model generated with Design Expert 18.0 has a good correlation between the experimental and predicted values because all of the points are very close to the line of perfect fit or line of unit slope [1, 2]. In addition, a residual investigation was carried out to validate the model's adequacy. The difference between the observed and

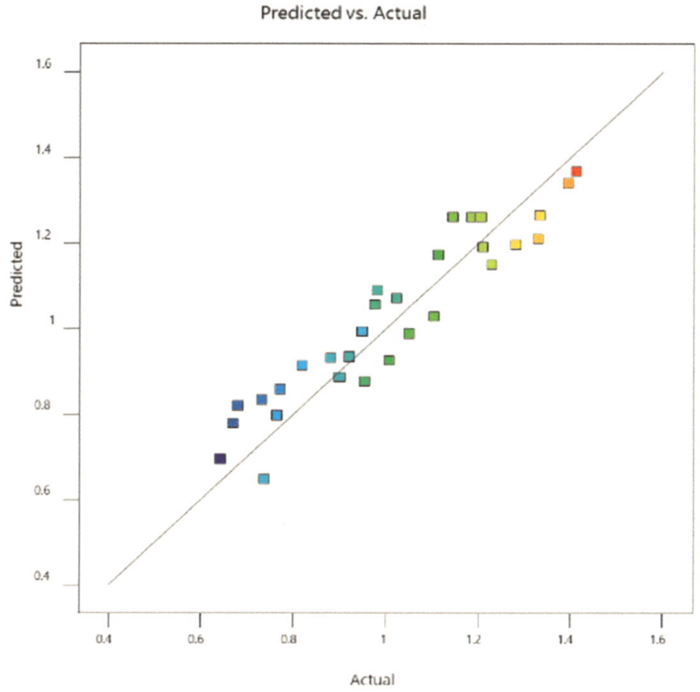

Figure 20.
Plot of experimental vs. predicted value WOH.

predicted responses is referred to as the residual. The normal probability plot of residuals was used to examine this analysis [2, 5]. The normal probability plot of the residuals shows that the errors are distributed normally in a straight line and are insignificant as shown in **Figure 21**.

5.3 Perturbation plot: width of heat affected zone

The perturbation plot shows the effect of all the parameters on a single plot. **Figure 22** shows a perturbation plot that compares the effect of all process parameters at the center point on bead width. WOH has been observed to increase as peak current (A) increases, and then decreases. The plot also shows that the WOH decreases as the base current (B) increases because no melting occurs during this stage. This plot shows that WOH increases as the pulse frequency (C) increases. The plot also shows that WOH increases as a pulse on time increases (D) and then decreases [2, 3].

5.4 Response surface plot: width of heat affected zone

The 3D surface plot and 2D contour effect developed by design expert 18.0 software represent the interaction effect between process parameters and WOH as shown in **Figures 23–28**. The coefficient of the linear effect of peak current and base current is positive as given in **Table 10**, WOH is increased as the value of the above parameter is increased as shown in **Figure 23a** of the 3D surface plot and **Figure 23b** of the contour plot. Peak current and base current are both rising at the same time as WOH,

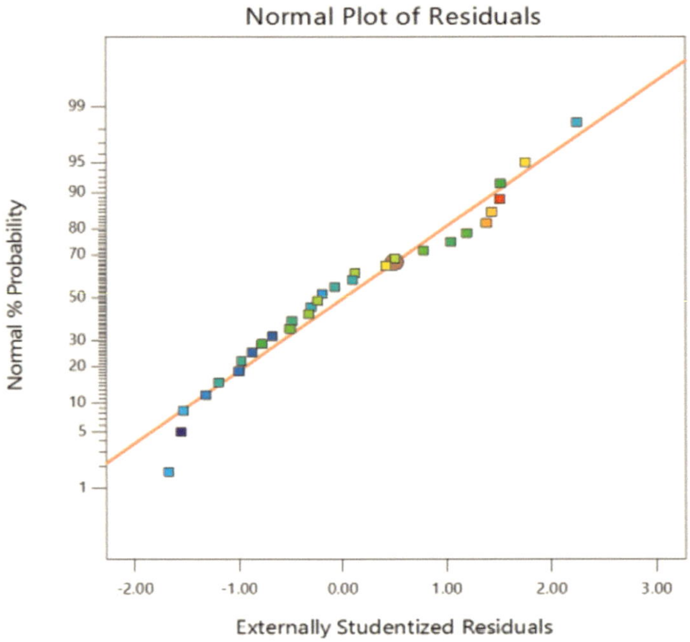

Figure 21.
Normal probability plot of residual WOH.

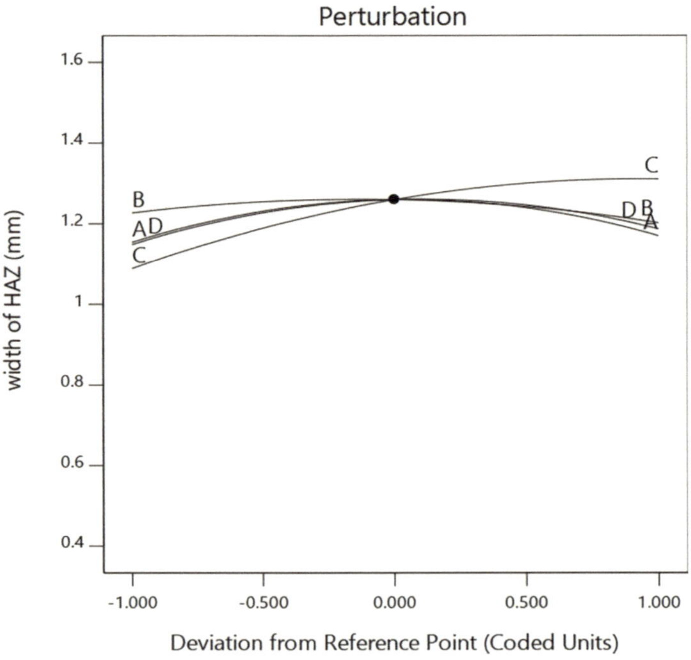

Figure 22.
Perturbation plot of WOH.

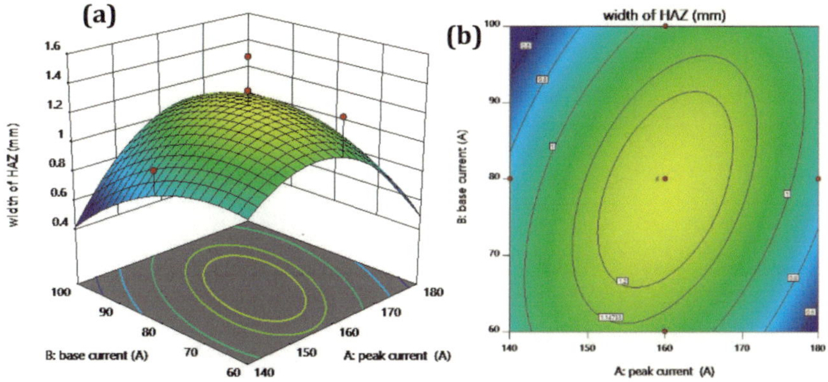

Figure 23.
Surface plot (a), contour plot (b) of the interaction effect AB on WOH.

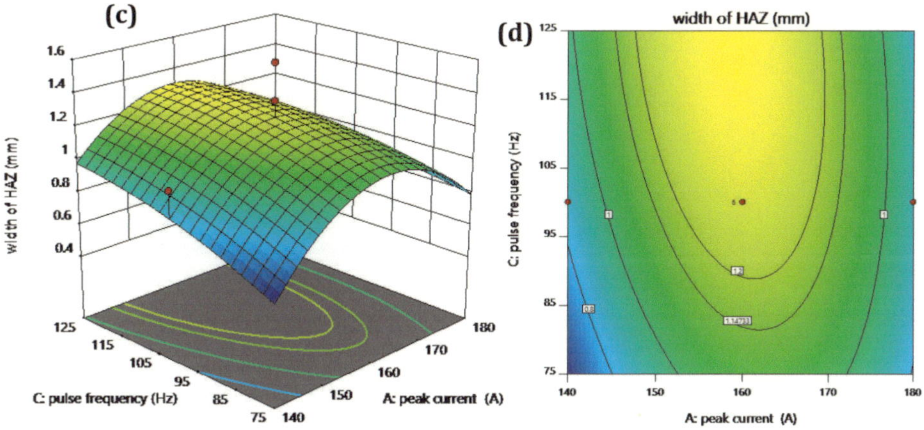

Figure 24.
Surface plot (c), contour plot (d) of the interaction effect AC on WOH.

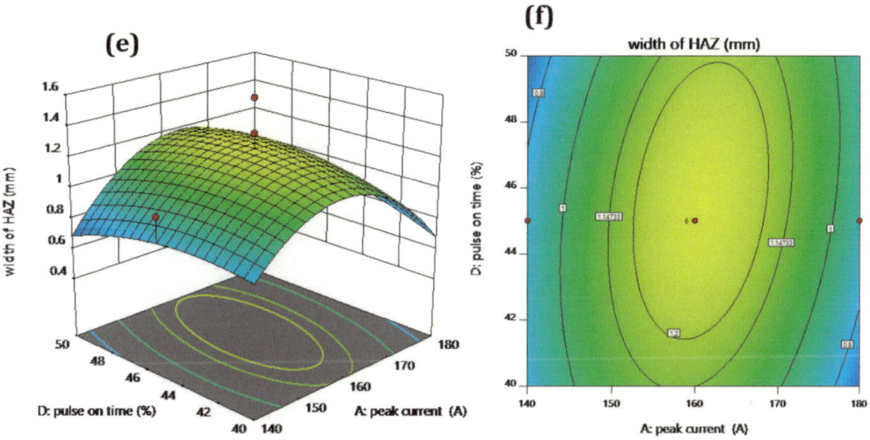

Figure 25.
Surface plot (e), contour plot (f) of the interaction effect AD on WOH.

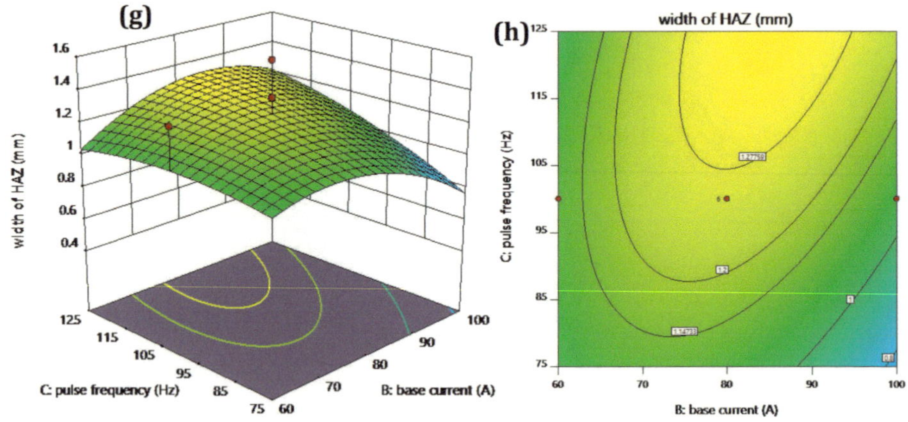

Figure 26.
Surface plot (g), contour plot (h) of the interaction effect BC on WOH.

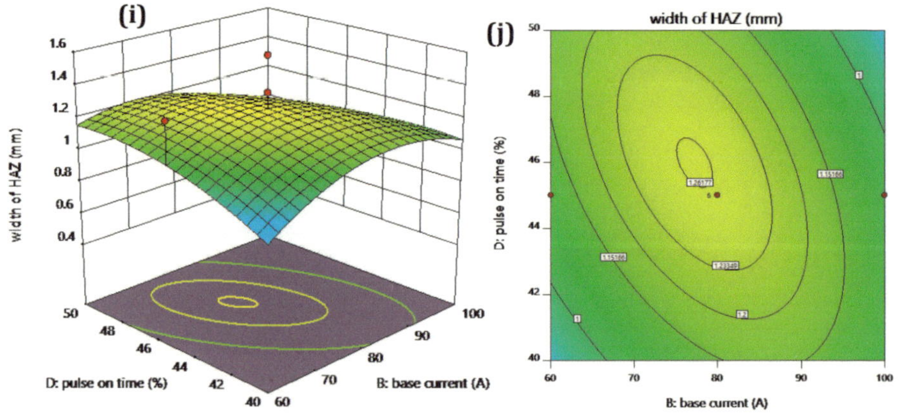

Figure 27
Surface plot (i), contour plot (j) of the interaction effect BD on WOH.

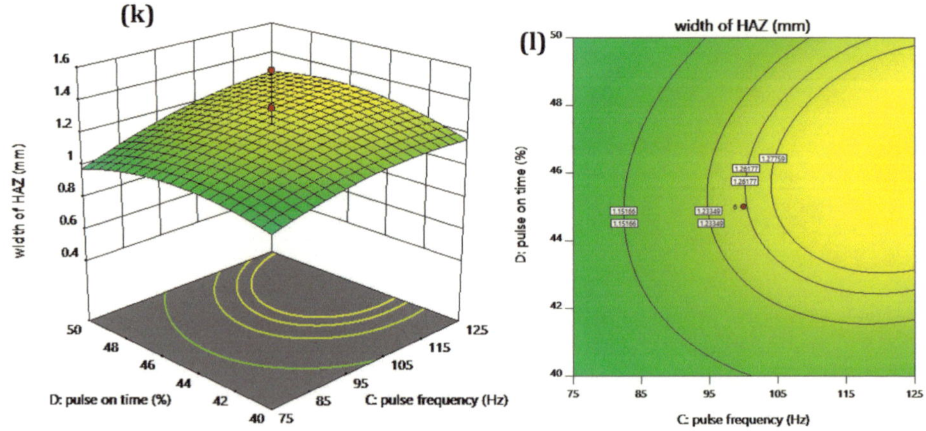

Figure 28.
Surface plot (k), contour plot (l) of the interaction effect CD on WOH.

reaching nearly 180 and 100 A, respectively, beyond which the value of WOH starts to drop. According to **Table 10**, the peak current and pulse frequency's coefficient of linear effects is both negative. WOH rises when the value of the aforementioned parameter rises, as demonstrated in **Figure 24c** and **d** of 3D surface plots and contour plots, respectively [2, 3]. The WOH declined beyond the peak current of 180 A and pulse frequency of 125 Hz respectively.

According to **Table 10**, the coefficient of the linear relationship between peak current and pulse on time is positive. WOH increases when the value of the aforementioned parameter increases, as demonstrated in **Figure 25e** and **f** of the 3D surface plot and the contour plot, respectively [2, 5]. WOH is increasing with simultaneously increasing in peak current and pulse on time to about 180 A and 50% respectively beyond which the value of WOH declines.

The coefficient of the linear effect of base current and pulse frequency is positive as given in **Table 10**, WOH is increased as the value of the above parameter is increased as shown in **Figure 26g** of the 3D surface plot and **Figure 26h** of the contour plot. WOH rises as base current and pulse frequency increase at the same time, peaking at roughly 100 A and 125 Hz, respectively, after which the value of WOH begins to decrease. According to **Table 10**, the coefficient of linear effects for base current and pulse on time is negative. When illustrated in **Figure 27i** of 3D surface plots and **Figure 27j** of contour plots, WOH increases as the value of the above parameter increases [1, 2]. The WOH declined beyond the base current 100 A and pulse on-time 50% respectively.

The coefficient linear effect of pulse frequency and pulse on time is positive as given in **Table 10**. WOH is increased as the value of the above parameter is increased as shown in **Figure 28k** of the 3D surface plot and **Figure 28l** of the contour plot [2, 3]. WOH is increasing with simultaneously increasing pulse frequency and pulse on time to about 125 Hz and 50% respectively beyond which the value of WOH declines.

6. Conclusion: Height of reinforcement

According to their greatest *F*-values in ANOVA **Table 4**, peak current and pulse frequency are the process variables that affect the HOR the most. The optimal conditions include a peak current of 160 A, a base current of 80 A, a pulse frequency of 100 Hz, a pulse on-time of 45%, and an optimal height of reinforcement that was projected to be 1.41 mm at this optimal condition. Experiments were conducted under these ideal conditions, as indicated in **Table 12**, to validate the projected optimum values. The experimental value of 1.45 mm matched the regression model's result very well. The constructed regression model is thus satisfied [2].

	Prediction	Experiment
Level	(160 A, 80 A, 100 Hz, 45%)	(160 A, 80 A, 100 Hz, 45%)
HOR mm	1.41	1.45

Table 12.
Confirmatory test: HOR.

	Prediction	Experiment
Level	(160 A, 80 A, 150 Hz, 45%)	(160 A, 80 A, 150 Hz, 45%)
DOH mm	1.32	1.37

Table 13.
Confirmatory test: DOH.

7. Conclusion: Depth of HAZ

Peak current and pulse frequency are the most significant process parameter that effects the DOH as indicated by their highest *F*-values given in the ANOVA **Table 7**. The optimal conditions are a peak current of 160 A, a base current of 80 A, a pulse frequency of 150 Hz, a pulse on-time of 45%, and an optimal HAZ depth of 1.32 mm under this optimal condition. To verify the projected optimum values, experiments were run under these ideal circumstances, as indicated in **Table 13**. The experimental value of 1.37 mm matched the regression model's result very well. The constructed regression model is therefore satisfied [2].

8. Conclusion: The width of HAZ

Peak current and pulse frequency are the most significant process parameter that effects the WOH as indicated by their highest *F*-values given in the ANOVA **Table 10**. The optimum conditions are the peak current of 160 A, the base current of 80 A, pulse frequency of 100 Hz, pulse on-time 45%, and optimum WoH at this optimum condition was predicted to be 1.26 mm. To validate the predicted optimum values, experiments were carried out at these optimum conditions [2]. The experimental value of 1.58 mm agreed closely with that obtained from the regression model as shown in **Table 14**. Therefore, the regression model developed is satisfied.

	Prediction	Experiment
Level	(160 A, 80 A, 100 Hz, 45%)	(160 A, 80 A, 100 Hz, 45%)
WOH mm	1.26	1.58

Table 14.
Confirmatory test: WOH.

Acknowledgements

The author made sincere thanks to all the technical staff of the ACMS laboratory of IIT Kanpur who directly and indirectly help in the experimental and analysis work. I would express my deep sense of gratitude to Dr. Shahnawaz Alam Sir and Dr. P.K. Bharti Sir for their valuable suggestions during this research work. I would also like to thank Chairman Sir (Shri Pranveer Singh Ji), Director Sir (Dr. Sanjeev Kumar Bhalla), and Shri Manmohan Shukla Ji (T&P) of Pranveer Singh Institute of Technology, Kanpur for their consistent encouragement and motivation.

Author details

Asif Ahmad[1*], Shahnawaz Alam[2] and Meenu Sharma[3]

1 Pranveer Singh Institute of Technology, Kanpur, Uttar Pradesh, India

2 Integral University, Lucknow, Uttar Pradesh, India

3 Government Polytechnic, Lucknow, Uttar Pradesh, India

*Address all correspondence to: aauptu2015@gmail.com

IntechOpen

© 2022 The Author(s). Licensee IntechOpen. This chapter is distributed under the terms of the Creative Commons Attribution License (http://creativecommons.org/licenses/by/3.0), which permits unrestricted use, distribution, and reproduction in any medium, provided the original work is properly cited.

References

[1] Ezekannagha CB, Ude CN, Onukwuli OD. Optimization of the methanolysis of lard oil in the production of biodiesel with response surface methodology. Egyptian Journal of Petroleum. 2017;**26**(4):1001-1011

[2] Ahmad A. Microstructure analysis and multi-objective optimization of pulsed TIG welding of 316/316L Austenite Stainless Steel. Handbook of Smart Materials, Technologies, and Devices: Applications of Industry 4.0. Hussain CM, Di Sia P. Cham: Springer International Publishing; 2020. pp. 1-33

[3] Song C, Dong S, He P, Yan S, Zhao X. Correlation of process parameters and porosity in laser welding of 7A52 aluminum alloy using response surface methodology. Procedia Manufacturing. 2019;**37**:294-298

[4] Varkey MJ, Sumesh A, Ramesh Kumar K. A Computational approach in optimizing process parameters influencing the heat input and depth of penetration of tungsten inert gas welding of Austenitic Stainless Steel (AISI 316L) using Response Surface Methodology. Materials Today: Proceedings. 2020;**24**: 1199-1209

[5] Iliyasu I, Bello JB, Oyedeji AN, Salami KA, Oyedeji EO. Response surface methodology for the optimization of the effect of fibre parameters on the physical and mechanical properties of deleb palm fibre reinforced epoxy composites. Scientific African. 2022;**16**:e01269

[6] Ahmad A, Alam S. Parametric optimization of TIG welding using response surface methodology. Materials Today: Proceedings. 2019;**18**: 3071-3079

Chapter 6

Analysis and Optimization of Process Parameters in Wire Electrical Discharge Machining Based on RSM: A Case Study

Aysun Sagbas

Abstract

In this book chapter a review and critical analysis on current research trends in wire electrical discharge machining (WEDM) and relation between different process parameters including pulse on time, pulse off time, servo voltage, peak current, dielectric flow rate, wire speed, wire tension on different process responses include material removal rate (MRR), surface roughness (Ra), sparking gap, wire lag and wire wear ration (WWR) and surface integrity factors was investigated. On the basis of critical evaluation of the available literature following conclusions are summarized. In addition, different modeling and optimization methods used in WEDM were discussed and a case study based on response surface method (RSM) including design of experiment (DoE) carried out to find optimal process parameters effect on surface roughness was conducted. In the final part of the present study was presented some recommendations about the trends for future WEDM researches.

Keywords: optimization, modeling, WEDM, RSM, DoE, surface quality

1. Introduction

The aim of this book chapter is to present some knowledge about the contributions of various researchers on WEDM process and to conduct an optimization approach named response surface methodology to determine the optimal process parameters. In addition, this book chapter is concluded by highlighting the optimal ranges of parameters in WEDM process for various materials and indicating the future research directions which will provide a reference to machine tool operators and manufacturing industries depending upon their demands. Moreover, this paper reviews and examines the various notable works in the field of WEDM and emphasis is made on optimization and modeling of machining parameters. The chapter also explains various advantages and disadvantages of different modeling and optimization methods used, and presents with some recommendations about trends for future WEDM researchers.

WEDM has a key role in unconventional machining method since it facilitates production of certain materials such as zirconium, titanium and intricate shapes. Wire

EDM is a thermo- electrical process which material is eroded by a series of sparks between the work piece and the wire electrode. The part and wire are immersed in a dielectric fluid which also acts as a coolant [1]. In EDM process, wire movement is monitored quantitively to obtain three dimensional shape. EMD has been known for more than a half century and used to manufacture high accuracy of the workpiece in machining processes and metal, tool, die, etc. industries. The development of the WEDM process was the result of seeking a technique to machine the electrodes used in EDM. In the end of the 1970s, computer numerical control (CNC) system was integrated with WEDM process. This integration was brought about a major evolution of the machining process. Moreover, the broad capabilities of the WEDM process were extensively exploited for any through hole machining owing to the wire, which has to pass through the part to be machined. It is probably the most exciting and diversified machine tool adopted for this industry in the last 50 years, and has various beneficial to use. In this process, there is no contact between electrode and work piece. Hence, materials of any hardness can be cut as long as they can conduct electricity. In addition, the wire does not touch the workpiece. So, physical pressure imparted on the workpiece is not exist, and amount of clamping pressure required to hold the workpiece is very low [2, 3]. Schematic diagram of WEDM process is given in **Figure 1**.

Recently, WEDM process has been widely used in manufacturing industry such as metals, alloys, sintered materials, cemented carbides, ceramics and silicon because of making micro-parts. These different systems support WEDM process which has remained as a competitive and reduced cost machining option fulfilling the demanding machining part requirements imposed by the short product development cycles and the growing cost pressures [4]. One of the most widely and commonly used and popular non-traditional material removal procedure which is currently often applied to manufacture components with complex shapes having great accuracy and precision is WEDM. Although, Wire-EDM uses a wire which acts as an electrode which is continuously traveling and is generally made up of thin brass, tungsten or copper, and is having a small diameter of 0.05–0.3 mm. Wire motion is regulated numerically to accomplish converted 3-dimensional shape and high precision of workpiece [5, 6]. Basic uses of wire electrical discharge machining incorporate extrusion tools and die, fixtures and gauges, models, airship and medical parts, and fabrication of stamping,

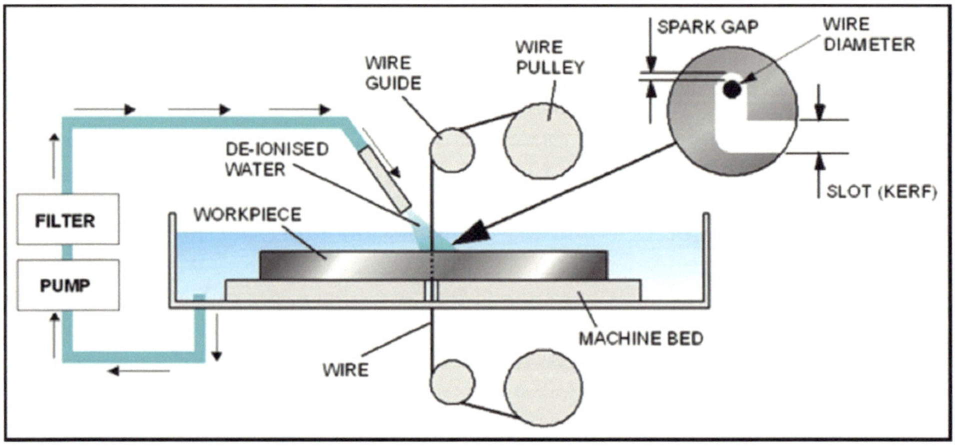

Figure 1.
Schematic diagram of WEDM process.

grinding wheel form tools [7]. Moreover, WEDM has been replacing other traditional machining operations in many industries throughout the world namely drilling, milling, grinding, turning, taper turning, etc. The setting for the various process parameters required in WEDM process play crucial role in achieving optimal performance. The main goals of WEDM manufacturers and users are to achieve a better stability and higher productivity of the WEDM process. Wire electrical discharge machining manufacturers and users emphasize on achievement of higher machining productivity with a desired accuracy and surface finish.

Response Surface Methodology is an important technique to use in designing, formulating, developing, and analyzing for different scientific studies and various industrial products. It is well known optimization method involves mathematical and statistical techniques. In RSM approach the objective is to optimize the response that is influenced by several input variables. Moreover, it is effictive and useful in the improvement and development of existing studies and products. Also, it can be applied to solve the optimization problems in different industries and it is commonly used Biological and Clinical Science, Social Science, Food Science, and Physical and Engineering Sciences. In many different manufacturing industries, one of the important issues is whether the system includes a maximum or a minimum or a saddle point, which has a wide important in industry. Therefore, RSM has been increasingly used in different industries. In addition, in recent years more emphasis has been placed by the chemical and processing field for finding optimal regions where there is an improvement in response instead of finding the optimum response [8]. The first aim for response surface method is to find the optimum response effected various input variables. When there are constraints on the design data, then the experimental design has to meet requirements of the constraints. The second purpose is to evaluate how the response changes in a given direction by adjusting the input and output variables [9, 10]. In generally, conventional data processing methods are not appropriate for investigating the process and product parameters. Many researchers have investigated the suitability of different empirical models to predict the changes in the quality parameters during different drying processes. Based on the function fitting technique, a response surface model in high dimensional space was fitted to show the relation between experiment inputs and output with minimum process knowledge. So, it could be used as an different alternative for conventional models like numerical simulation during optimization with a reduced computational cost and time according to the other various optimization techniques. On the other hand, RSM-based models are only accurate for predicting the relationship between a limited number of input and output parameters. Box-Behnken and central composite design (CCD) commonly used in many research, both have its advantages [11–13].

In this study a predictive model (RS model) is developed and applied to optimize WEDM machining parameters using RSM approach. Experiments are carried out to test the the validity and accuracy of model and satisfactory results are obtained. The methodology described here is expected to be highly beneficial to manufacturing industries such as aerospace, chemistry, textile, automobile and tool making, etc. industries.

2. Literature review: process modeling and optimization

Due to large number of process parameters and responses lots of researchers have attempted to improve the process capability. Some researchers have used different optimization techniques such as Taguchi technique, gray relational analysis

(GRA), design of experiment, artificial neural network (ANN) modeling, desirability approach and evolutionary algorithm. Lots of authors tried to model this process using the Taguchi method and response surface methodology approach [5, 14–18] which utilized response surface methodology coupled with gray-Taguchi technique. Further, Lin [19] have combined Taguchi method with the GRA to optimize the micro milling EDM performance. Similarly, hybrid approach of Taguchi gray has been used by Rajyalakshmi and Ramaiah [20] for multiple performance optimization of WEDM machined Inconel 825. In contrast to WEDM performance evaluation, Sharma [2] have used one factor at a time approach to investigate the effect of various WEDM control parameters on performance characteristics. Except conventional techniques of optimization, some evolutionary algorithm has been in literature such as genetic algorithm (GA), artificial bee colony (ABC), particle swarm optimization (PSO), teaching learning-based optimization (TLBO) and differential evolution (DE). These algorithms provide a global optimum solution instead of local optimum solutions. The parametric settings named optimal solution is found out based on optimization techniques like VIKOR based Harmony search algorithm and desirability function approach to get perfect surface finish during electrical discharge coating and electrical discharge machining of AISI 1040 stainless steel and Nitinol respectively [5]. Further, statistical models have been developed by Kuppan et al. [21] to determine the relationship between EDM output responses and control parameters using response surface methodology. Similarly, Ramakrishnan and Karunamoorthy [22] have developed the mathematical model based on Box and Hunter central composite design to determine the effect of control parameters on EDM performance characteristics. Further, Ramakrishnan and Karunamoorthy [23] presented an Artificial Neural

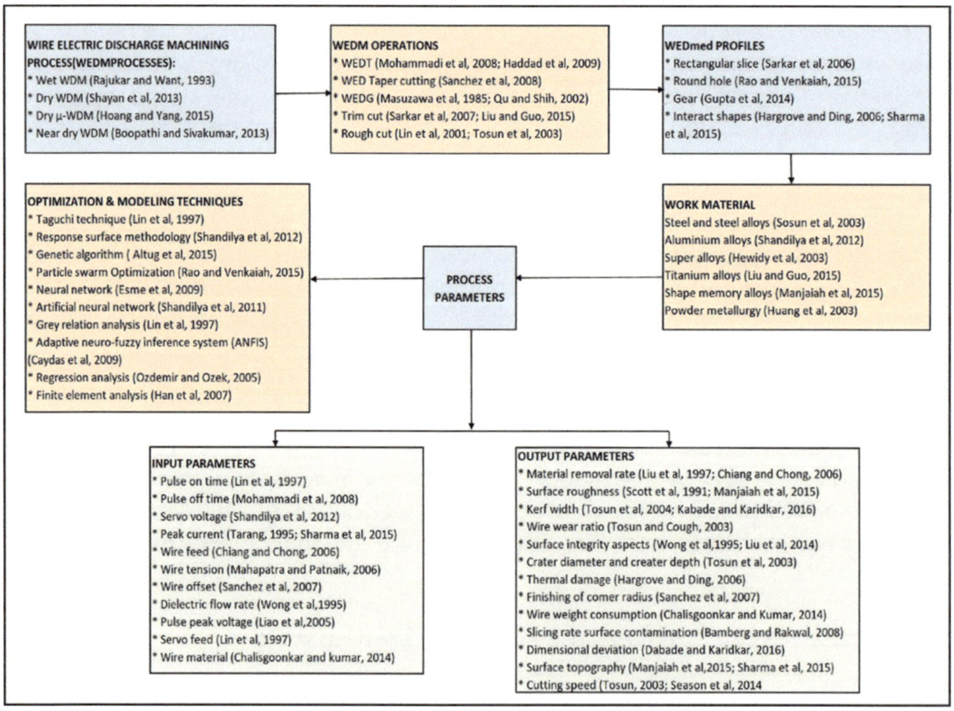

Figure 2.
Comparative study of published research work on WEDM [24].

Network model to predict the WEDM performance of Inconel 718 alloy. **Figure 2** shows a brief outline of past research works.

In cutting operation, WEDM primarily employed either for trim cut [25–27] or rough cut [28, 29]. To the best knowledge of authors, this technique can be successfully employed for machining of steel and steel alloys [30–33] aluminum and aluminum alloys, titanium and its alloys [27, 34] super alloys [35, 36] metal matrix composites [37, 38] green compact manufactured by powder metallurgy [39]. Investigations into the influences of machining input parameters on the performance of WEDM have been widely reported [25, 40–42]. Several attempts have been made to develop mathematical model of the WEDM process [39, 43–47]. In these works, productivity of the process and the surface roughness of the machined work piece are examined as measures of the process performance. Neural network models on material removal rate in EDM has been studied by Tsai and Wang [48] whereas Lee and Li [49] investigated on effects of process parameters in EDM using tungsten carbide as work material. Qu et al. [50] have, through examination of literature, concluded that research has not been directed towards EDM applications in the area of newly developed engineering materials and the boundaries that limit the material removal rate (MRR). Scott, Boyina et al. [43] used a factorial design method, to determine the optimal combination of control parameters in WEDM considering the measures of machining performance as metal removal rate and the surface finish. Tarng and Chung [51] carried out a neural network model to estimate cutting speed and surface finish using input parameters such as pulse duration, pulse interval, peak current, open, servo reference voltage, circuit voltage, electric capacitance and table speed. Trezise [52] presents that essential limits on machining accuracy are dimensional consistency of the wire and the positional accuracy of the work table. Sarkar et al. [25] studied the WEDM of titanium aluminide. They also attempted to develop an appropriate machining strategy for a maximum process yield criterion. A feed forward back propagation neural network was used to model the machining process. Ali [53] investigated on the effect and optimization of machining parameters on the surface roughness in the WEDM process of AlCu-TiC-Si P/M composite. The optimal machining parameters were obtained by using Taguchi experimental design method. The variation of MRR and surface roughness with machining parameters is mathematically modeled by using non-linear regression analysis method. Patil and Brahmankar [54] examined the effect of various input parameters such as pulse on time, pulse off time, ignition pulse current, wire speed, wire tension and flushing pressure on cutting speed and surface finish of Al/SiCp by using Taguchi methods. Shandilya et al. [38] concluded that to achieve higher value of the average cutting speed, lower value of voltage and higher value of pulse-off time should be used during WEDC of SiCp/6061 Al MMC. In the most recent work, They studied the effect of input process parameters on surface surface roughness during WEDM of SiCp/6061 Al MMC. There are some researches that used traditional approach for modeling WEDM like Tarng [51] which utilized feed forward neural network to model and simulated annealing (SA) algorithm is then applied to the neural network to solve the optimal cutting parameters problem. Other one of them is Lin et al. [19] which used Taguchi method with fuzzy logic for modeling and optimization. In addition, Huang [30] studied Wire-EDM based on Gray relational and statistical analyses. Furthermore Kuriakose et al. [55] applied data mining approach. Yuan et al. [56] used incorporating prior model into Gaussian processes regression for WEDM process modeling. Also Caydas, et al. [32] used neuro-fuzzy inference system (ANFIS) to model this process. Besides Cheng et al. [42] utilized a neural network integrated simulated annealing approach for optimizing WEDM. Kapil K. at all [57] investigated the cutting rate and recast layer thickness while designing the servo feed, pulse on-time, servo voltage, and pulse off-time with the Box-Benkhen design of RSM. Kumar et all [45] the Box-Benkhen design of response surface methodology based and machine learning algorithm was applied for the WEDM process, to simultaneously optimize SR, MRR of CP-Ti G2 [58].

3. Case study

In this study, desired surface roughness is obtained based on four input parameters by creating an experimental model of AISI 4030 steel and using response surface methodology. This study has shown that RSM model presented here has overcome WEDM complex that results in satisfactory surface quality characteristics. Each experimental test was conducted twice and averaged as R_a mean values to acquire database with high confidence. Furthermore, experiments were designed using the method that was introduced by Box and Hunter [59]. The experimental runs were performed as per the central composite design which is a type of response surface methodology designs. Response surface methodology has been used to plan and analyze the experiments. CCD was used in order to fit the second order response in surface as well as in optimization methods for finding relation between various individual input parameters and reactions. **Table 1** demonstrates coded value and actual values of individual parameters, and **Table 2** shows machining conditions in WEDM process.

Parameter	Levels				
	−2	−1	0	1	2
Open circuit voltage (V)	60	120	180	240	300
Wire speed (m/min)	2	4	6	8	10
Dielectric flushing pressure (kg/cm^2)	6	9	11	14	16
Pulse duration (ns)	10	37	50	725	900

Table 1.
Experimental factors and factor levels.

Workpiece	AISI 4340
Electrode	CuZn37
Workpiece dimensions (mm)	150 × 150 × 10
Table feed rate (mm/min)	8.2
Pulse interval time (s)	18
Wire diameter (mm)	0.25
Wire tensile strength (N/mm^2)	900
Cut-off length (mm)	0.8

Table 2.
Machining conditions in WEDM process.

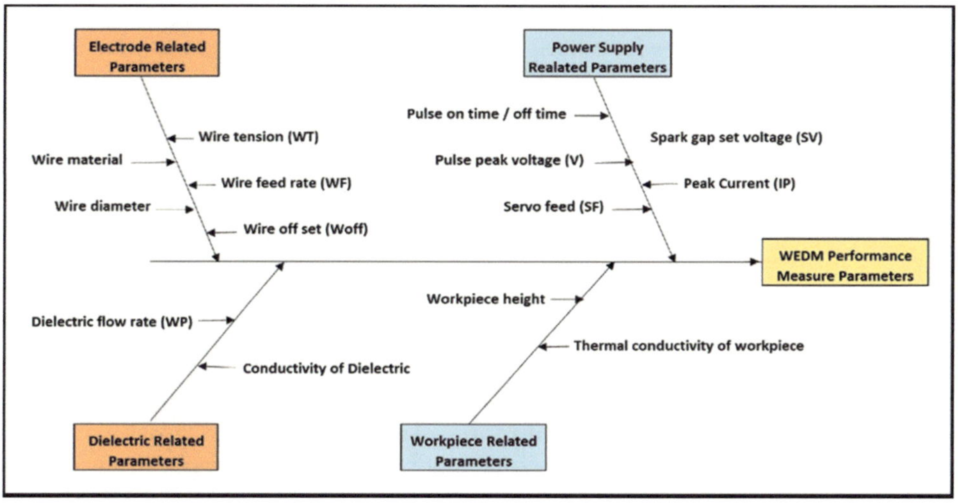

Figure 3.
Cause and effect diagram for WEDM process parameter [24].

3.1 Process parameters used in study

The most important performance measures in WEDM are metal removal rate, surface finish, and cutting width. They depend on machining parameters like discharge current, pulse duration, pulse frequency, wire speed, wire tension and dielectric flow rate. In WEDM process, it is seen that, input process parameters such as pulse-on time, pulse-off time, servo voltage, peak current, wire feed rate, wire tension, wire offset, water pressure, servo feed, wire material are having significant influence on process parameters named surface roughness, kerf width, material removal rate, wire wear rate, surface integrity aspects, etc. [3–9, 16–18]. The various input process parameters of WEDM and their inter-relationship is presented using Ishikawa's cause-effect diagram shown in **Figure 3**.

In this study, open circuit voltage, wire speed, and dielectric flushing pressure were selected as input parameters and surface roughness was selected as output parameter.

3.2 Statistical analysis and modeling using RSM

In Response surface methodology approach responses of interest is influence by several variables and in which the objective is to optimize these responses [59, 60]. In this method the effects of the noise factors have been considered. In addition, statistical optimization model can overcome the limitation of classical methods to obtain the optimum process conditions. Predictive model (RS model), which is an analytical function, in predicting response surface is formulated as following polynomial function:

$$Y = a_0 + \sum_{i=1}^{n} a_i X_i + \sum_{i=1}^{n} \sum_{j=1}^{n} a_{ij} X_i X_j + \ldots + \varepsilon \tag{1}$$

where Y is the desired response, a_0 is constant, a_i and aij represent the coefficients linear, quadratic terms, respectively, X_1 reveals the coded variables corresponding to the studied machining parameters, input variables, n is the number of the model parameters, ε is the random error.

In this study, a predictive model was developed to reach low surface roughness in terms of cutting parameters for milling operations. RSM design was tested with 30 data sets of central composite design of experiment. Surface roughness (R_a) measurements were made by using Phynix TR–100 portable surface roughness tester. Surface roughness measurements were made by using Phynix TR–100 portable surface roughness tester. To identify the significant factors for WEDM process, analysis of variance (**Table 3**) was employed by using Design Expert software.

This table demonstrates that the terms in the model have a significant effect on the responses. It is found that, the open circuit voltage has the most dominant effect on the surface roughness followed by the pulse duration and wire speed respectively. Goodness of fit for model generated by experimental data was evaluated and analyzed based on ANOVA. This includes the tests for significance of model, their coefficients and lack of fit model adequacy. ANOVA is used to create, access and analyze the experimental test data and goodness of fit model is generated afterwards. Through the backward elimination process, the final quadratic models of response equation in terms of coded factors are as follows:

$$Y = 114 + 17.73X_1 + 3.27X_2 + 1.27X_3 + 3.47X_4 + 15.85X_1X_{2-}2.65X_1X_{3-}3.30X_1X_4 + 0.91X_2X_3 + 8.019X_2X_{4-}1.36X_3X_4 - 0.75X_1^2 - 0.111X_2^2 + 0.15X_3^2 - 0.44X_4^2 \quad (2)$$

When the regression model above is examined, change in wire speed has significant impact on surface roughness. In this context, as wire speed increases, increase in surface roughness is observed. There is a strong linear relationship between the surface roughness and open circuit voltage, whereas there is a weak relationship between dielectric flushing pressure between surface roughness. It proves the complex influence of the adopted input variables on the analyzed value of the surface roughness. This model includes experimental test data that shows models importance, coefficients, and inadequacy in model fit.

In this study experimental surface roughness values were compared with surface roughness predicted values of the RS model. It was observed that the prediction of surface roughness closely agrees with that of the experimental values. Moreover, measured surface roughness has been correelated well with the predicted surface roughness values. It was also found that the RS model for the predicted values generates an average best fit percentage error of 6.83%. The involvement of process factors on surface roughness for WEDM process was analyzed with the help of surface graphs for the selected process factor combinations are presented in **Figures 4–15. Figures 4** and **5** represent interaction graphs wire speed between open circuit voltage and open circuit voltage between dielectric flushing pressure graphs respectively.

Figure 4 shows that at lower wire speed the effect of open circuit voltage on surface roughness is statistically insignificant. However, at higher wire speed the effect of open circuit voltage on surface roughness is important and statistically significant. Similarly, with the rise in wire speed value, the lower value of open circuit results better surface roughness.

It is demonstrated that in **Figure 5**, at lower or higher open circuit value the effect of dielectric flushing pressure on surface roughness is very poor. **Figures 6** and **7** represent interaction graphs wire speed between pulse duration and dielectric flushing between pulse duration graphs respectively.

As seen in **Figure 6** at lower pulse duration the effect of wire speed on surface roughness is not statistically important, whereas, at higher pulse duration level this effect partly more significant. When the wire speed increases at low pulse duration

Source	Sum of squares	df	Mean Square	F-value	p-value
Model	13,239.59	14	945.68	37.98	< 0.0001
X_1- Open circuit volt	7211.01	1	7211.01	289.58	< 0.0001
X_2- Wire speed	256.96	1	256.96	10.32	0.0058
X_3- Dielectric flush. pr.	38.43	1	38.43	1.54	0.2332
X_4- Pulse duration	288.36	1	288.36	11.58	0.0039
X_1X_2	4021.78	1	4021.78	161.51	<0.0001
X_1X_3	112.20	1	112.20	4.51	0.0508
X_1X_4	174.17	1	174.17	6.99	0.0184
X_2X_3	13.27	1	13.27	0.5328	0.4767
X_2X_4	1072.40	1	1072.40	43.07	< 0.0001
X_3X_4	29.40	1	29.40	1.18	0.2943
X_1^2	15.76	1	15.76	0.6329	0.4387
X_2^2	0.3427	1	0.3427	0.0138	0.9082
X_3^2	0.6232	1	0.6232	0.0250	0.8764
X_4^2	5.54	1	5.54	0.2223	0.6441
Residual	373.52	15	24.90		
Lack of fit	318.50	10	31.85	2.89	< 0.0001
Pure error	55.02	5	11.00		< 0.0001
Cor total	13,613.11	29			< 0.0001

Table 3.
The analysis of variance (ANOVA) on the performance of surface roughness.

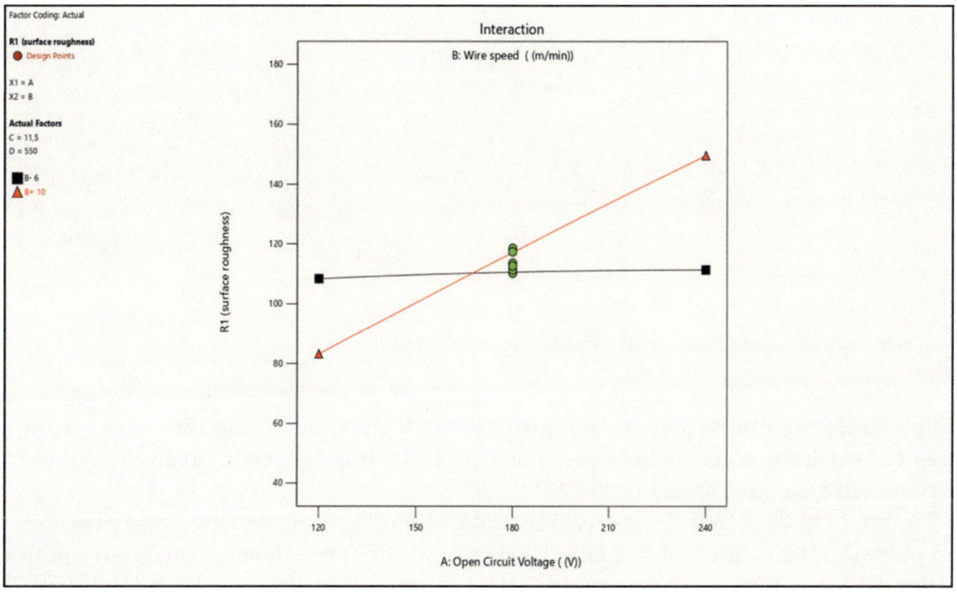

Figure 4.
Interaction graph wire speed and open circuit voltage on surface roughness.

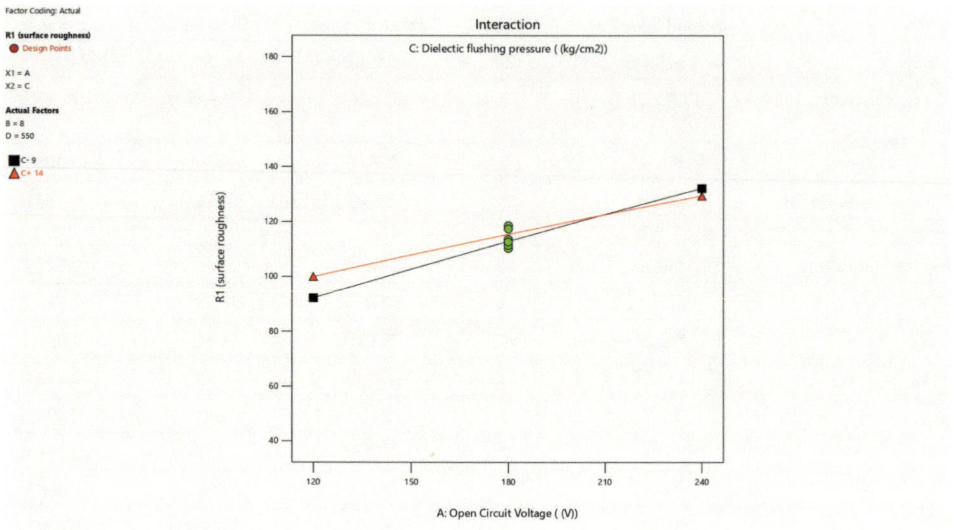

Figure 5.
Interaction graph open circuit voltage and dielectric flushing pressure on surface roughness.

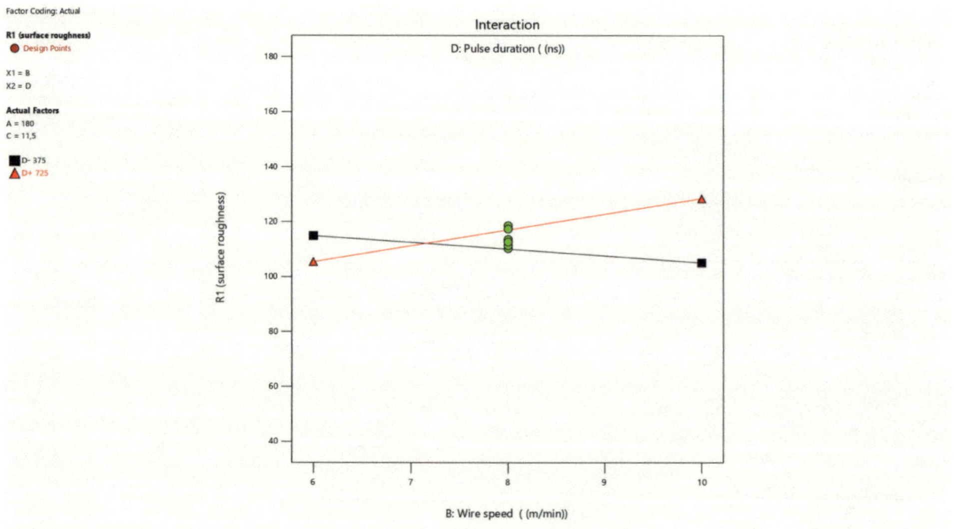

Figure 6.
Interaction graph wire speed between pulse duration on surface roughness.

value surface roughness increases. **Figures** 7 and **8** show that the interaction graphs dielectric flushing pressure between pulse duration and dielectric flushing pressure between wire speed graphs respectively.

Figure 7 exhibits that the increasing and decreasing dielectric flushing pressure and pulse duration values it has no statistically significant effect on surface roughness. In **Figures 8** and **9** they are represented the interaction graphs wire speed between dielectric flushing pressure and open circuit voltage between pulse duration graphs respectively.

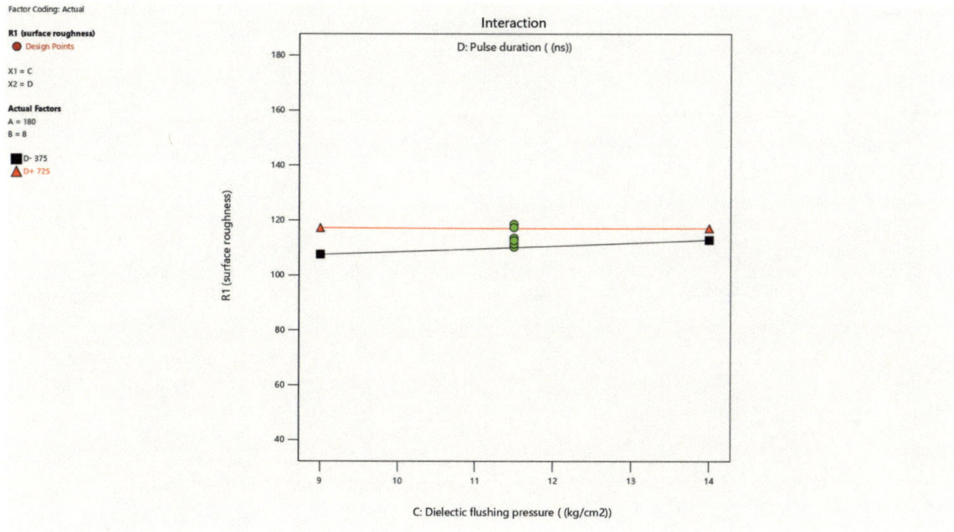

Figure 7.
Interaction graph dielectric flushing pressure between pulse duration on surface roughness.

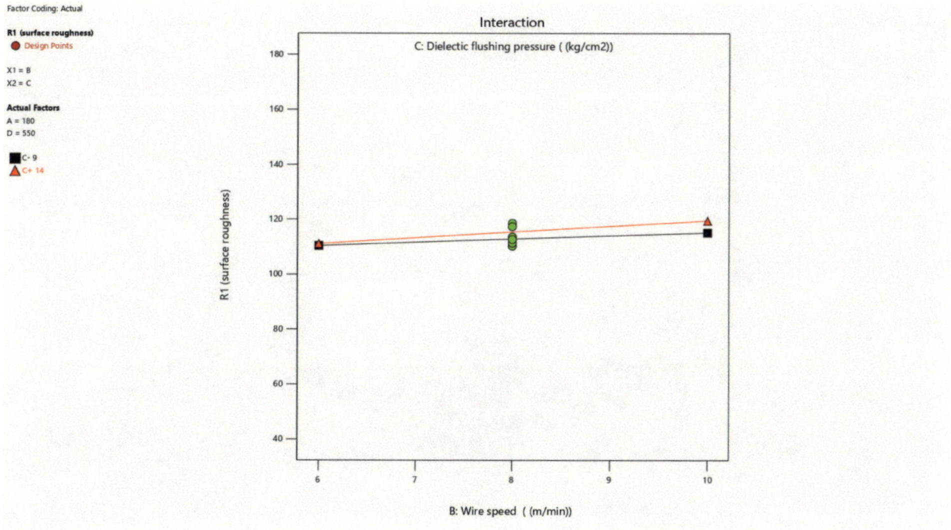

Figure 8.
Interaction graph wire speed and dielectric flushing pressure on surface roughness.

As seen in **Figure 8** in case of increasing or decreasing wire speed and dielectric values the magnitude of the surface roughness does not change significantly in level of significant 5%.

Figure 9 demonstrates the variation of pulse duration and open circuit voltage concerning the surface roughness. It can be causes that increasing surface roughness with the increase in pulse duration when the open circuit increases. Similar trends were observed that lower pulse duration the effect of open circuit on surface roughness is poorer statistically. **Figures 10–15** represent 3D contour plot graphs input and

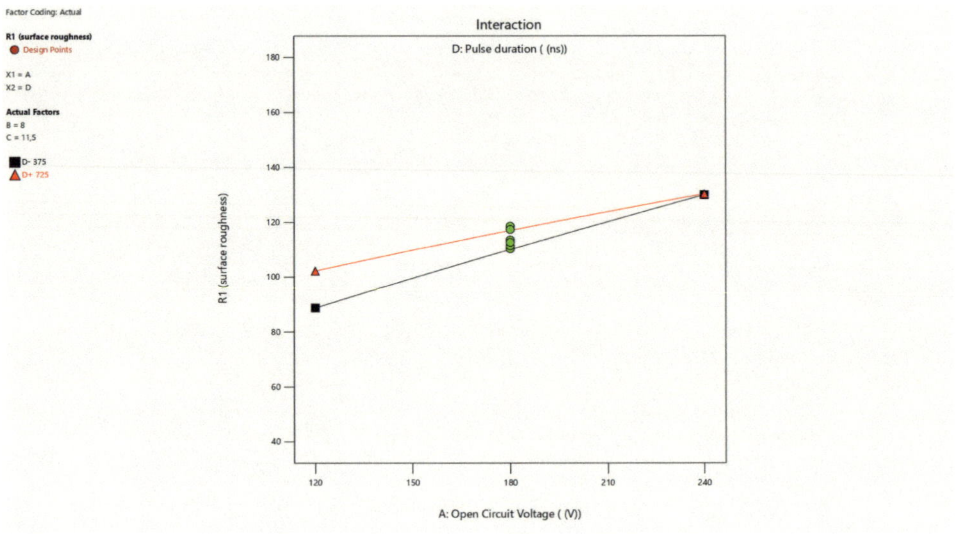

Figure 9.
Interaction graph open circuit voltage and pulse duration on surface roughness.

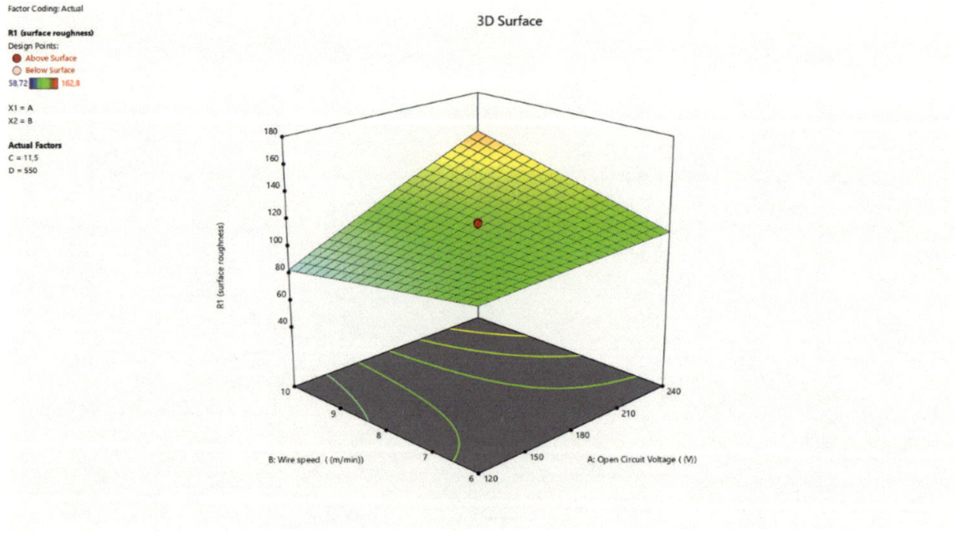

Figure 10.
Contour plot graph wire speed and open circuit voltage on surface roughness.

response variables. **Figures 10** and **11** represent 3D contour plot graphs wire speed between open circuit voltage and open circuit voltage between pulse duration graphs respectively.

As seen in **Figure 10** it was identified that the higher wire speed with the lower open circuit value results better surface roughness. Moreover, when wire speed is increased in case of lower open circuit voltage the value of surface roughness is very poor.

Figure 11 exhibits the surface roughness decreases with an decrease in open-circuit voltage and increase pulse duration. Also, higher pulse duration with lower open circuit voltage the value of surface roughness is minimum. **Figures 12** and **13**

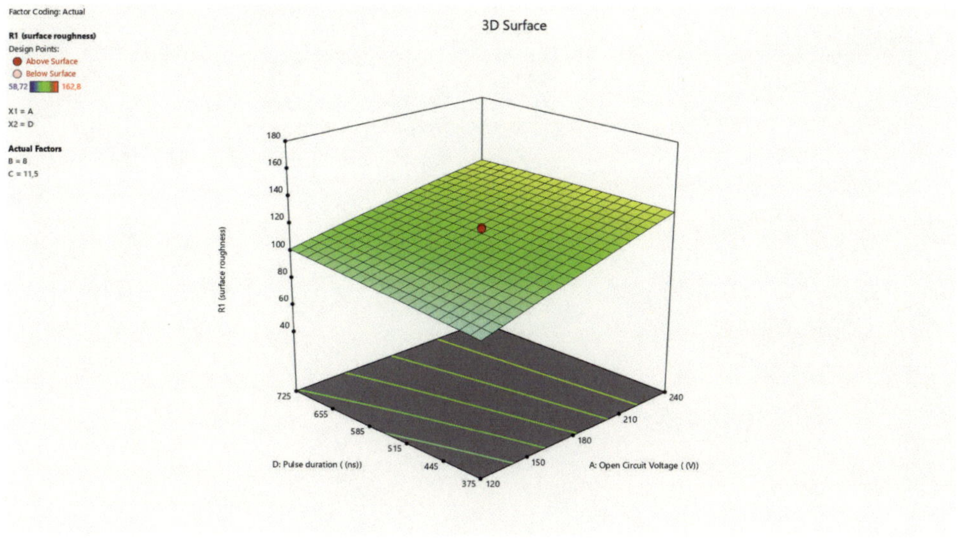

Figure 11.
Contour plot graph pulse duration and open circuit voltage on surface roughness.

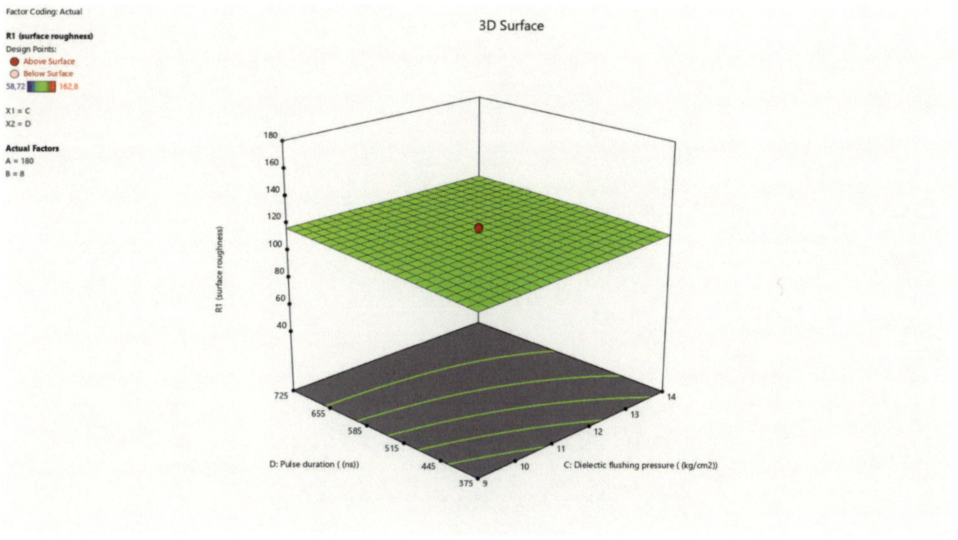

Figure 12.
Contour plot graph pulse duration between dielectric flushing pressure on surface roughness.

represent pulse duration between dielectric flushing pressure and wire speed between dielectric flushing pressure 3D contour plot graphs respectively.

It was noticed from **Figure 12** the lower magnitude of dielectric flushing pressure in case of lower wire speed value decreases surface roughness. **Figures 13** and **14** represent dielectric flushing pressure between wire speed and pulse duration between wire speed 3D contour plot graphs respectively.

As seen in **Figure 13** if dielectric flushing pressure is increased when wire speed is decreased, decreasing in surface roughness is observed. It was perceived that when the higher value of dielectric flushing pressure in case of lower wire speed the value of surface roughness is very poor.

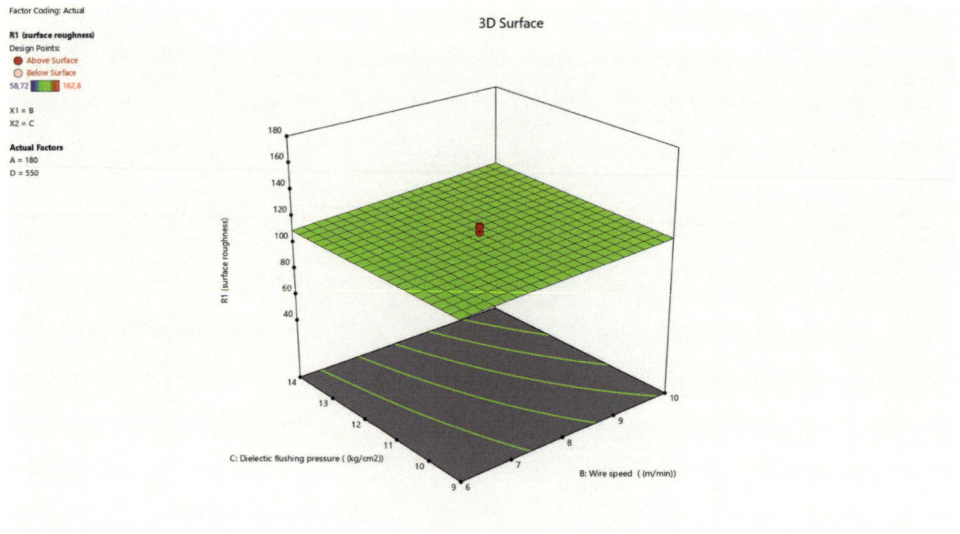

Figure 13.
Contour plot graph wire speed and dielectric flushing pressure on surface roughness.

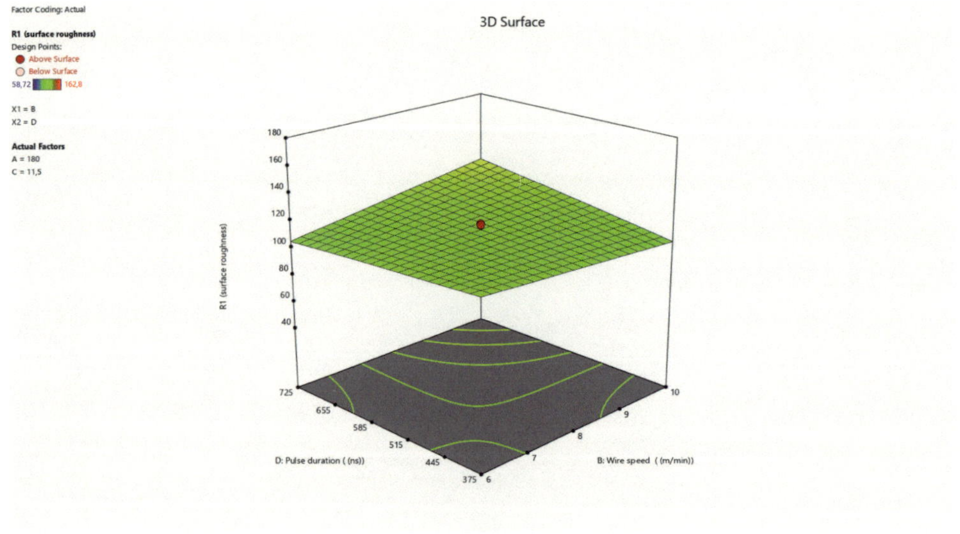

Figure 14.
Contour plot graph wire speed and pulse duration on surface roughness.

Figure 14 shows that at low wire speed with low pulse duration surface roughness value increases. The low values of dielectric when higher pulse duration may cause surface roughness also reduces. In addition, it is predictable from **Figure 14**, the combination of pulse duration and dielectric in lower range gives a good surface finish. **Figure 15** represents dielectric flushing pressure between open circuit voltage 3D contour plot graph.

From **Figure 15** with the increase in wire speed in case of lower open circuit voltage, it can be obtained lower surface roughness. Furthermore, increasing of high of dielectric flushing pressure in case of lower dielectric leads to decreasing surface roughness in WEDM process.

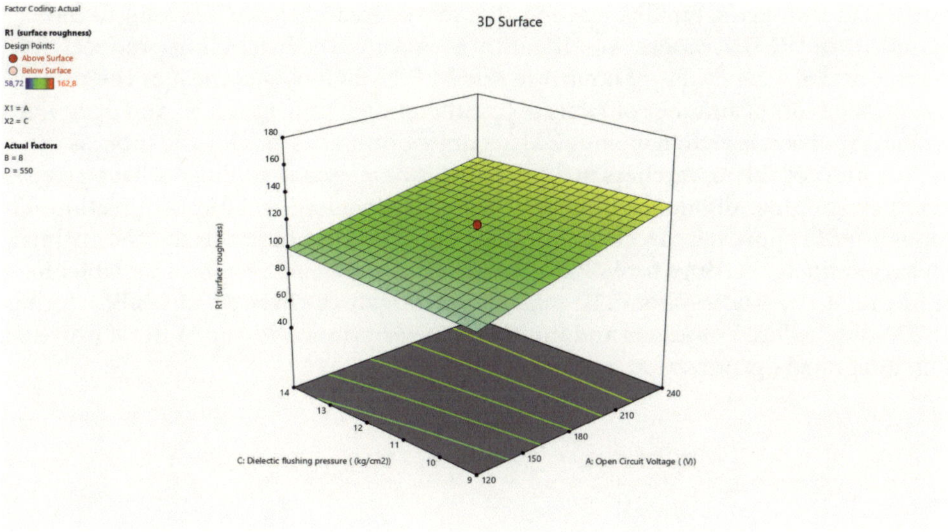

Figure 15.
Contour plots graph dielectric flushing pressure and open circuit voltage on surface roughness.

It is evident that, a predicted optimum surface roughness obtained from the response surface and contour plots by using RSM, a pulse duration of 346 (ns), open circuit voltage of 142 (V), wire speed of 8 (m/min) and dielectric flushing pressure of 12 (kg/cm^2) is 2.63(μm). The objective of developed model is to establish the quantitative relationship between output and input control parameters. It is seen that, RSM model proposed here in has resolved the complex of WEDM process that results in satisfactory surface quality characteristics. Hence, the experimental results confirm that the developed model predicts effectively and the optimal process parameters significantly improve in the WEDM process. As a result, predicted optimum surface roughness was acquired. Results from the adopted design of the experiment, where the explanatory variables were determined independently of each other, is a desirable feature because it indicates the uniqueness of the prediction.

4. Conclusions and future scope

This study mainly focuses on the development of empirical model of AISI 4340 steel in WEDM process to obtain the desired surface roughness in terms of four prominent input parameters using response surface methodology. In WEDM process, optimization of the response variable is very important and essential problem for various scientific studies and manufacturing industries. Because WEDM is an expensive production process and widely used in many manufacturing process such as aerospace, chemistry, textile, automobile and tool making industries. The essential purpose of the WEDM process is to achieve an accuracy and efficiency in production process. Several researchers have studied with different methods to improve the surface quality and increase the material removal rate of the WEDM process. However, the problem of selecting the cutting parameters in the WEDM process is not completely solved. Still there is lack of information about different WEDM wire types. Hence, more research should be done about comparing different inputs on different responses. Finally, it seems that more researches can be strength the capabilities of WEDM process significantly to improve the machining productivity, accuracy and efficiency. From literature review it is obvious that most of the researchers examined lots of number of process parameters at a time to model and optimize various responses, which may not yield accurate optimal values for the process. Further, most of the researchers include both academics and applicants have given the importance to individual and multi -response modeling and its optimization. The proposed RSM approach can effectively assist engineers in determining the optimal process parameter settings for WEDM process for individual response variable. In the future, many studies should be made to investigate the process capability during WEDM of powdered products and multi response optimization on WEDM process by using integrated optimization methods.

Author details

Aysun Sagbas
Corlu Engineering Faculty, Namik Kemal University, Tekirdag, Turkey

*Address all correspondence to: asagbas@nku.edu.tr

IntechOpen

© 2022 The Author(s). Licensee IntechOpen. This chapter is distributed under the terms of the Creative Commons Attribution License (http://creativecommons.org/licenses/by/3.0), which permits unrestricted use, distribution, and reproduction in any medium, provided the original work is properly cited.

References

[1] Kuriakose S, Shunmugam MS. Characteristics of wire-electro discharge machined Ti6Al4V surface. Materials Letters. 2004;**58**:2231-2237

[2] Sharma-Khanna R, Gupta RD. WEDM process variables investigation for HSL by response surface methodology and genetic algorithm. Engineering Science and Technology, an International Journal. 2015;**18**:171-177

[3] Zain AM, Haron H, Sharif S. Application of GA to optimize cutting conditions for minimizing surface roughness in end milling machining process. Expert Systems with Applications. 2010;**7**:4650-4659

[4] Nourbakhsh F, Rajurkar KP, Malshe AP, Cao J. Wire electrodischarge machining of titanium alloy. Procedia CIRP. 2013;**5**:13-18

[5] Mahapatra SS, Patnaik A. Optimization of wire electrical discharge machining (WEDM) process parameters using Taguchi method. The International Journal of Advanced Manufacturing Technology. 2007;**34**:9-10

[6] Williams R, Rajurkar KP. Study of wire electrical discharge machining surface characteristics. Journal of Materials Processing Technology. 1991;**28**:486-493

[7] Kasdekar DK, Parashar V. Performance of Box-Behnken and prediction model are compared with AL-Nano MMC to maximize the MRR. Materials Today: Proceedings. 2017;**4**:3173-3181

[8] Rajesha S, Sharma AK, Kumar P. On electro discharge machining of Inconel 718 with hollow tool. Journal of Materials Engineering and Performance. 2012;**21**(6):882-891

[9] Huang CA, Hsu CC, Kuo HH. The surface characteristics of P/M high-speed steel (ASP 23) multi-cut with wire electrical discharge machine (WEDM). Journal of Materials Processing Technology. 2003;**140**:298-302

[10] Puri AB, Bhattacharyya B. An analysis and optimisation of the geometrical inaccuracy due to wire lag phenomenon in WEDM. International Journal of Machine Tools & Manufacture. 2003;**43**:151-159

[11] Sarkar S, Mitra S, Bhattacharyya B. Parametric optimisation of wire electrical discharge machining of γ titanium aluminide alloy through an artificial neural network model. International Journal of Advanced Manufacturing Technology. 2006;**27**:501-508

[12] Hsu HS. Preliminary study of material removal in electrical-discharge machining of SiC/Al. Journal of Materials Processing Technology. 1997;**63**:813-818

[13] Lin J, Wang KS, Yan BH, Tarng YS. Optimization of the electrical discharge machining process based on the Taguchi method with fuzzy logics. Journal of Materials Processing Technology. 2000;**102**:48-55

[14] Hewidy MS, El-Taweel TA, El-Safty MF. Modelling the machining parameters of wire electrical discharge machining of Inconel 601 using RSM. Journal of Materials Processing Technology. 2005;**169**:328-336

[15] Kung K-Y, Chiang K-T. Modeling and analysis of machinability evaluation in the wire electrcal

discharge machining (WEDM) process of aluminum oxide-based ceramic. Materials and Manufacturing Processes. 2008;**23**:241-250

[16] Liao YS, Huang JT, Su HCA. study on the machining-parameters optimization of wire electrical discharge machining. Journal of Materials Processing Technology. 1997;**71**:487-493

[17] Liao YS, Huang JT, Chen YH. A study to achieve a fine surface finish in Wire-EDM. Journal of Materials Processing Technology. 2004;**149**:165-171

[18] Huang TS, Hsieh SF, Chen SL, Lin MH, Ou SF, Chang WT. Surface modification of TiNi-based shape memory alloys by dry electrical discharge machining. Journal of Materials Processing Technology. 2015;**221**:279-284

[19] Rajyalakshmi G, Venkata RP. Multiple process parameter optimization of wire electrical discharge machining on Inconel 825 using Taguchi grey relational analysis. International Journal of Advanced Manufacturing Technology. 2013;**69**(5-8):1249-1262

[20] Boopathi S, Sivakumar K. Experimental investigation and parameter optimization of near dry wire cut electrical discharge machining using multiobjective evolutionary algorithm. International Journal of Advanced Manufacturing Technology. 2013;**67**:2639-2655

[21] Kuppan P, Rajadurai A, Narayanan S. Influence of EDM process parameters in deep hole drilling of Inconel 718. The International Journal of Advanced Manufacturing Technology. 2008;**38**(1-2):74-84

[22] Ramakrishnan R, Karunamoorthy L. Multi response optimization of wire EDM operations using robust design of experiment. International Journal of Advanced Manufacturing Technology. 2006;**29**:105-112

[23] Ramakrishnan R, Karunamoorthy L. Modeling and multi-response optimization of Inconel 718 on machining of CNC WEDM process. Journal of Materials Processing Technology. 2008;**207**(1):343-349

[24] Singh B, Mısra JP. A critical review of wire electric discharge machining. DAAAM International Scientific Book. 2016;**23**:249-266

[25] Matsuo T, Oshima E. Investigation on the optimum content and machining condition for wire edm of zirconia ceramics. CIRP Annals. 1992;**41**:231-234

[26] Liu JF, Guo YB. Process capability of wire-edm of niti shape memory alloy at main cut and trim cut modes. Procedia Manufacturing. 2015;**1**:904-914

[27] Chalisgaonkar R, Kumar J. Investigation of the machining parameters and integrity of the work and wire surfaces after finish cut WEDM of commercially pure titanium. Journal of Brazial Society Mechanical Science Engineering. 2016;**38**(3):883-911

[28] Tosun N, Cogun C, Pihtili H. The effect of cutting parameters on wire crater sizes in Wire EDM. The International Journal of Advanced Manufacturing Technology. 2003;**21**:857-865

[29] Boopathi S. Experimental comparative study of near-dry wire cut electrical discharge machining (WEDM). European Journal of Scientific Research. 2012;**75**:472-481

[30] Huang JT, Liao YS. Optimization of machin- ing parameters of Wire-EDM based on grey relational and statistical

analyses. International Journal of Production Research. 2003;**41**:1707-1720

[31] Spedding TA, Wang ZQ. Parametric optimization and surface characterization of wire electrical discharge machining process. Precision Engineering. 1997;**20**(1):5-15

[32] Caydas U, Hascalik A, Ekici S. An adaptive neuro-fuzzy inference system (ANFIS) model for wire-EDM. Expert Systems with Applications. 2009;**36**:6135-6139

[33] Sagbas A, Gürtuna F, Polat U. Comparison of ANN and RSM modeling approaches for WEDM process optimization. Materials Testing. 2021;**63**(4):386-392

[34] Manjaiah M, Narendranath S, Basavarajappa S, Gaitonde VN. Effect of electrode material in wire electro discharge machining characteristics of Ti50Ni50 shape memory alloy. Precision Engineering. 2015;**41**:68-77

[35] Goswami A, Kumar J. Investigation of surface integrity, material removal rate and wire wear ratio for WEDM of Nimonic 80A alloy using GRA and Taguchi method. Engineering Science and Technology, an International Journal. 2014;**17**:173-184

[36] Dabade UA, Karidkar SS. Analysis of response variables in WEDM of inconel 718 using taguchi technique. Procedia CIRP. 2016;**41**:886-891

[37] Manna A, Bhattacharyya B. Taguchi and gauss elimination method: A dual response approach for parametric optimization of cnc wire cut EDM of PRAlSiCMMC. International Journal of Advanced Manufacturing Technology. 2006;**28**:67-75

[38] Shandilya P, Jain PK, Misra JP. Experimental investigation during wire electric discharge cutting of SiCp/6061 aluminium metal matrix composite. Annals of DAAAM for 2010. In: Proceedings of the 21st International DAAAM Symposium. Germany: DAAM; 2010;**21**(1):1091-1092

[39] Hatami S, Navid MS, Nyborg L. Surface preparation of powder metallurgical tool steels by means of wire electrical discharge machining. Metallurgical and Materials Transactions. 2012;**43**:3215-3226

[40] Rajurkar KP, Wang WM. Thermal modeling and on-line monitoring of wire-ED. Journal of Materials Processing Technology. 1993;**38**(1-2):417-430

[41] Sone T, Masui K. Application of ion nitriding to wireelectrical-discharge-machined blanking dies. Materials Science and Engineering. 1991;**140**:486-493

[42] Chen HC, Lin JC, Yang YK, Tsai CH. Optimization of wire electrical discharge machining for pure tungsten using a neural network integrated simulated annealing approach. Expert Systems with Applications. 2007;**37**:7147-7153

[43] Scott DS, Boyina KP, Rajurkar L. Analysis and optimisation of parameter combination in wire electrical discharge machining. International Journal of Production Research. 1991;**29**(11):2189-2207

[44] Indurkhya G, Rajurkar KP. Artificial neural network approach in modeling of edm process. In: Proceedings of artificial neural networks in engineering (ANNIE' 92) Conference. Missouri, U.S.A: St.Louis; 1992. pp. 845-890

[45] Kumar SA, Sharma R, Gupta A, Gujral R. Investigation of biocompatible implant material through WEDM process using RSM modeling hybrid with the

machine learning algorithm. Indian Academy of Sciences. 2021;**46**(148)

[46] Sharma P, Chakradhar D, Narendranath S. Measurement of WEDM performance characteristics of aero-engine alloy using RSM-based TLBO algorithm. Measurement. 2021;**179**:1-13

[47] Pramanick A, Mandal A, Dey PP, Das PK. Comparative analysis for the prediction of WEDM responses for machining spark plasma sintered boron carbide ceramic sample by RSM and ANFIS. Materials Today: Proceedings. 2021;**41**:1089-1095

[48] Tsai HC, Huang FY, Lee LC. Examination of wire electrical discharge machining of Al2O3/6061Al composites. International Journal of Machine Tools & Manufacture. 2005;**45**:251-259

[49] Lee SH, Li X. Study of the effect of machining parameters on the machining characteristics in electrical discharge machining of tungsten carbide. Journal of Materials Processing Technology. 2001;**115**:334-358

[50] Qu J, Shih AJ, Scattergood RO. Development of the cylindrical wire electrical discharge machining process: Part I: Concept, design, and material removal rate. ASME Journal of Manufacturing Science and Engineering. 2002;**124**:702-707

[51] Tarng YS, Ma SC, Chung LK. Determination of optimal cutting parameters in wire electrical discharge machining. International Journal of Machine Tools and Manufacture. 1995;**35**(12):693-170

[52] Trezise KE. A Physicist's view of wire EDM. Proceedings of the International Conference on Machine Tool Design and Research. 1982;**23**:413-419

[53] Ali MY, Mustafizul KAN, Adesta EYT, Ismail AF, Abdullah AA, Idris MN. Comparative study of conventional and micro WEDM based on machining of meso/micro sized spur gear. International Journal of Precision Engineering and Manufacturing. 2010;**11**:779-784

[54] Patil NG, Brahmankar PK. Some investigations into wire electrodischarge machining performance of Al/SiCp composites. International Journal of Machining and Machinability of Materials. 2006;**1**(4):412-431

[55] Kuriakose S, Mohan K, Shunmugam MS. Data mining applied to wire-EDM process. Journal of Materials Processing Technology. 2003;**142**:182-189

[56] Yuan J, Liu CL, Liu X, Wang K, Yu T. Incorporating prior model into Gaussian processes regression for WEDM process modeling. Expert Systems with Applications. 2009;**36**(4):8084-8092

[57] Kapil K, Goyal NS, Gupta RD, Singh G, Rani D, Banga HK, et al. A soft computing-based analysis of cutting rate and recast layer thickness for AZ31 alloy on WEDM using RSM-MOPSO. Materials. 2022;**15**(635):1-22

[58] Montgomery DC. Design and Analysis of Experiments. 7th ed. New Jersey: John Wiley & Sons; 2009. p. 964

[59] Garg MP, Jain A, Bhushan G. Multi-objective optimization of process parameters in wire electric discharge machining of Ti-6-2-4-2 Alloy. Arabian Journal for Science and Engineering. 2013;**39**:1465-1476

[60] Myers RH, Montgomery DC. Response Surface Methodology: Process and Product Optimization using Designed Experiments. Second ed. New York: Wiley; 2002. pp. 8084-8092

Chapter 7

Optimization of Baker's Yeast Production on Grape Juice Using Response Surface Methodology

Sawsan Mahmood, Ali Ali, Ayhem Darwesh and Wissam Zam

Abstract

The purpose of this study is to complete as an example the fermentation conditions allowing the production of *Saccharomyces cerevisiae* yeast biomass in large quantities using the juice as the same carbon source. Determination of the best of five factors affects the production of dry biomass by baker's yeast. The optimal value of the five factors affecting the process of biomass production by the baker's sourdough was determined. The experimental design was performed using CCD (Central Composite Experimental Design), and the response surface methodology method was used to determine the best possible amount of production of yeast and has reached (41.44 g/L) after 12 hours of fermentation, under the following optimal conditions (temperature (30.11°C), pH (4.75), sugar concentration (158.36 g/L), the ratio of carbon to nitrogen (an essential nutrient for yeast growth) that is (11.9), and initial concentration of yeasts (2.5 g/L). Three kinematic models (Monod, Verhulst, and Tessier) were also selected for the purpose of studying the kinetic performance of *S. cerevisiae* yeast, and the best results were obtained based on the Verhulst model. The Leudeking Piret model has also been successfully used to estimate substrate during fermentation.

Keywords: *Saccharomyces cerevisiae*, response surface methodology, kinetic models, assumption, statistics

1. Introduction

Fermentation is one of the oldest methods used by humans since ancient times to preserve food and improve its organoleptic properties. More than 5000 fermented foods and beverages are produced worldwide, from alcohol, beer, and vinegar to cheese, yogurt, sourdough bread, olives, sausages, kimchi, and soybean paste [1].

Fermentation is simply the biochemical transformation of raw materials which is supported by the synthesis and stabilization of bacteria which convert sugars into simple acids, alcohols, and carbon dioxide to improve the flavor, texture, and aroma of processing and extend the shelf life of fermented products. Goods. During fermentation, many secondary metabolites including vitamins, antioxidants, and bioactive compounds are formed by the microbial community, contributing to the nutritional and nutraceutical value of the final product [2].

There has also been a rapid and significant development in fermentation technologies in recent years after understanding the bio-physiology of microorganisms and controlling it. Among this biology is the yeast, which has received more attention after recent developments in understanding its physiology [3].

Yeasts are micro-organisms, single-celled, unicellular eukaryotes. Their shapes and structure differ from one species to another. They are spherical or oval in shape and their dimensions range between 5 and 30 μm in length and 3–10 m in width. The yeast multiplies quickly and grows well in the contained environment. On sugars where they multiply by budding or by division [4, 5]. Yeasts play vital roles in food biotechnology, especially in fermented products [6].

S. cerevisiae yeast is the most important type of yeast due to its use in many industrial fields. It is used in the production of food, bread, pastries, ethyl alcohol, beer, wine, and as well as in the production of single-cell protein and a number of medicinal foods [7, 8].

S. cerevisiae yeast is considered to be the most important product of biotechnology due to its widespread use in the industrial field [9].

S. cerevisiae biomass is produced by using bioreactors that contribute to controlling growth conditions and the production is carried out according to batch or fed-batch fermentation system [10].

Baker's yeast industrially relies on a variety of disciplines, including variations of different generations, times and stages of aeration, differentiation of bioreactors, and control of the final stage of cultivation [11]. It is an aerobic process based on the expansion of cells from pure culture to larger bioreactors by increasing the volume at each stage of expression in the sugar medium [12].

Commercial bread yeast comes in three forms: Pressed yeast that is sold in the form of pressed briquettes or cubes wrapped with wax paper or cellophane, and its shelf life does not exceed one week from the date of its production due to the speed of its corruption. Active dry yeast is sold in airtight containers and needs to be reactivated before use, its cells are about 8–10% moisture and its shelf life ranges from six months to a year depending on the storage temperature. The instant dry yeast contains 4–5% moisture and its shelf life reaches more than a year and is added to the dough directly without the need for revitalization [13].

The global yeast market is estimated to be valued at USD 3.9 billion in 2020 and is projected to reach USD 6.1 billion by 2025 [14].

Molasses is the most used raw material in the production of Baker's yeast, it may be sourced from sugar beet or sugar cane, and contains about 50–55% of fermentable sugars, some vitamins and minerals that are important in cell proliferation, also any substance containing fermentable sugars can be used such as the date and grape juices [15].

In the last years, the price of molasses has increased because of their use in other industrial applications such as animal feeding or bioethanol production [16], thus rendering the evaluation of new substrates for yeast biomass propagation a trending topic for biomass producers' research. New assayed substrates include molasses mixtures with corn steep liquor (20:80), different agricultural waste products [17], and other possibilities such as date juice or agricultural waste sources, also called wood molasses that can be substrate only for yeast species capable of using xylose as a carbon source [18].

In this research, the possibility of using grape juice to produce a good yield from the yeast was studied in this study. Grape juice was chosen because it has a chemical composition similar to the chemical composition of molasses in terms of its good content of hexane-sugars and its richness with many important nutrients for the growth of yeast cells, in addition to the fact that grape cultivation is spread in various parts of the world, including Syria, which is one of the grape-producing countries.

During the last war period, Syria was exposed to difficult economic conditions and the suspension of the work of the only sugar factory in the country, and this was accompanied by the suspension of the yeast factory and the tendency to import yeast. So, the researchers went to study the possibility of an alternative or additional option for molasses that supports yeast production, and this is in line with the researchers' interest. In different parts of the world studying the possibility of using available raw materials to support biotechnology industries and finding many options or alternatives that support any vital industry. The Syrian Arab Republic is the richest country in the Middle East in the cultivated varieties of grapes, and the number of varieties is about 100 varieties spread across the country where the most important varieties are spread, which are four varieties that represent 85 percent of the total grape production (Zaini 15%, Baladi 20%, Salti 20%, and Heloani 30%). The main objective of the present work is to study the optimization of *S. cerevisiae* biomass production, using grape juice as the only source of carbon, as grape juice is a good source of carbon and many important nutrients for the growth of yeast, and it has a chemical composition close to the chemical composition of molasses [19].

The efforts of many researchers are directed toward improving various biological manufacturing processes [20], including fermentation processes, with the aim to determine the best conditions for the production of the required product, as well as with the aim to solve the problems that may face the required manufacturing process, reaching the highest possible production of the final product and reducing the costs of the manufacturing process as possible [21, 22].

Several statistical experimental design methods have been used to optimize biological processes [23, 24].

These methods, including the central composite experimental design (CCD), are characterized by reducing the number of experiments required, reducing financial and energy costs, reducing the time required, as well as reducing the reagents and materials required during work [25, 26].

The central composite experimental design (CCD) is one of the methods that contributed to the improvement of a number of biological processes such as the production of antibiotics, enzymes, organic acids, and ethanol [27, 28].

The study was conducted by selecting the best for five measurements (temperature, initial pH, sugar content in the juice, carbon to nitrogen ratio, and primary yeast) in order to get high yields of yeast using optimization with the surface response methodology method, we use grape juice as carbon source for cell growth and produce *S. cerevisiae* at high performance, and finally predict the biomass production process with three kinetic models.

2. Material and methods

2.1 Origin and reactivation of the yeast *S. cerevisiae*

The yeast used in this study is a commercial yeast from the sigma company, it is a dry powder form of *S. cerevisiae* (ATCC20408/S288c). This yeast needs to be reactivated before use with a suitable nutrient medium Yeast Peptone Glucose Agar

(YPGA) consisting of 20 g/L agar, 10 g/L yeast extract, 10 g/L glucose, 10 g/L peptones with a pH 6, with incubated at 30°C for 24 h.

2.2 Preparation of grape juice

The Baladi grape (**Figure 1**) was chosen and it is one of the varieties available in Syria. Its production reaches 20% of the grape production. It is a local variety that is distinguished by the size of its large clusters and has a single conical shape, and the grains are spherical in shape, with a large size, a yellowish-white color, and a thin crust in a light pink color. The pulp is flaky, has a good taste, and has a distinctive flavor, one of the late-ripening varieties, and it is one of the famous and luxurious table varieties, suitable for remote transportation and long winter storage.

The grape is obtained from local markets. The grape berries were removed from their clusters and cleaned and washed with warm water. The juice was extracted by breaking and pressing in a doubly folded cloth, then the juice was pasteurized at 85°C for 3 minutes.

2.3 Preparation of culture medium based on grape juice and inoculums

The juice resulting from the above preparation was supplemented with mineral salts: 0.44 g of magnesium sulphate, 12.70 g of urea, and 5.30 g of ammonium sulphate. Finally, the medium was placed in 250 mL deltas at a volume of 100 mL per well and sterilized at 120° C for 20 min. The preculture was obtained by inoculating two colonies of *Saccharomyces cerevisiae* yeast in 250 mL flasks containing 100 mL of juice as mentioned above. Pre-cultures were incubated at 30°C for 3 h and then used as inoculum for potassium biomass production [29].

Figure 1.
The Baladi grape.

2.4 Statistical design of experiments

2.4.1 Factor selection and organization of experiments

Five independent variables were selected (temperature, initial pH, concentration of sugars in grape juice, the ratio of carbon to nitrogen, and initial concentration of yeasts).

In a previous study, carried out by Naser and Abdelrahman [30], with the aim of determining the optimal conditions for producing baker's yeast using sugar cane molasses and achieving the best yield and lowest production cost, the best results were obtained when using the concentration of sugars within the range (14–18) %, Yeast inoculum level 2 to 3 g/L, agitation speed between 150 r.p.m. and 200 r.p.m., adding (40–50) % urea and ammonium sulfate at pH = 5.

In another study by Muhammad et al. [31], the baker's yeast production process was improved and the effect of various physical and chemical factors on the production of yeast cells was evaluated. The optimal conditions were determined to obtain the maximum possible growth of yeast cells at a concentration of sugars equal to 100 g/L, the agitation speed at 150 r.p.m., at pH = 4.5, and T = 28°C.

Optimization of baker's yeast production using date juice as the sole carbon source using the response surface methodology method has been studied by Ali et al. [32] and the study showed the success of using date juice in obtaining a good yield of the yeast biomass at the initial conditions of the fermentation process (sugar concentration 70.93 g/L, temperature 32.9°C and pH 5.35).

A study carried out by Taleb et al. [33] showed that the use of ammonium sulfate and urea as a source of nitrogen during the production of break's yeast by (50–50) % contributed to improving production yield by more than 36%, and adding thiamine vitamin at a concentration of 0.6 had a positive role in improving production by more than 6%.

A study by Sokchea et al. [34] indicated that the best amount of biomass for yeast is obtained when the ratio between the concentration of glucose and nitrogen (C/N) used during the fermentation process is equal to 10.

After reviewing previous studies, the lower and upper levels of studied variables were selected, **Table 1** shows the lower and upper levels of studied variables.

A program Minitab 19 software was used to optimize the Baker's yeast production The CCD matrix is composed of a complete factorial design, 32 cube points, eight center points in a cube, 10 axial points, and four center points in axial design variable at a distance of α = 2.366 and two-level factorial. Each experiment was carried out twice and the average value is used.

variables	Lower level (−1)	Upper level (+1)
X1 = Temperature (°C)	25	35
X2 = Initial pH	3	6
X3 = Concentration of sugars (g/L)	100	200
X4 = The ratio of carbon to nitrogen	8:1	15:1
X5 = Initial concentration of yeasts (g/L)	2	3

Table 1.
The lower and upper levels of studied variables.

2.4.2 Effect estimation

The real values X have been calculated according to Eq. (1).

$$X = \left(x - x_{\circ}\right)/\Delta x \quad (1)$$

Where, X is the coded value for the independent variable, x, is the natural value, x_0, is the natural value at the center point, and ΔX is the step change value (the half of the interval (−1 + 1)).

Regression Equation in Uncoded Units:

$$\begin{aligned} Yi = {} & \beta_0 + \beta_1X_1 + \beta_2X_2 + \beta_3X_3 + \beta_4X_4 + \beta_5X_5 + \beta_{11}{X_1}^2 + \beta_{22}{X_2}^2 + \beta_{33}{X_3}^2 \\ & + \beta_{44}{X_4}^2 + \beta_{55}{X_5}^2 + \beta_{12}X_1X_2 + \beta_{13}X_1X_3 + \beta_{14}X_1X_4 + \beta_{15}X_1X_5 \\ & + \beta_{23}X_2X_3 + \beta_{24}X_2X_4 + \beta_{25}X_2X_5 + \beta_{34}X_3X_4 + \beta_{35}X_3X_5 + \beta_{45}X_4X_5 \end{aligned} \quad (2)$$

Yi is the predicted response (the Biomass production (g/L). The calculation of the effect of each variable and the establishment of a correlation between the response Yi and the variables X were performed using a Minitab 19 Statistical Software (Minitab, Inc., State College, PA, USA) [32].

2.5 Statistical analysis

The statistical analysis was performed using (ANOVA), in order to validate the square model regression. It included the following parameters: coefficient of determination R^2; Fisher test (F); p-value and Student test (t); and the statistical significance test level was set at (probability <0.05) [32].

2.6 Validation of biomass production in optimum medium

After completing the optimization of the production of baker's yeast in grape juice, the optimum values obtained, and representative of the fermentation conditions were confirmed by conducting an experiment.

The experiment was carried out on 250 mL shake flasks and the agitation speed was 200 r.p.m. To do this, 100 mL of grape juice was seeded with 11 mL of the yeast pre-culture and the pH of the medium was adjusted to the obtained value of 4.75. Shake flasks were sterilized at 120°C for 20 min and incubated at 30°C (optimum Value) for 12 h.

2.7 Analytical methods

2.7.1 Determination of total reducing sugars

1 ml of the sample is taken after filtering it and placed in a glass tube, then 98% sulfuric acid and 0.6 mL of 5% (w/v) phenol were added and mixed well after which it is left at room temperature for 30 minutes, the absorbance is measured using a spectrophotometer (Analytik Jena- specord 200uv-vis spec.) at a wavelength of 490 nm, the concentration of the reducing sugar is calculated depending on the calibration curve, which was formed between different concentrations of standard

solutions of glucose and between the absorbance values corresponding to each concentration [35].

2.7.2 Determination of biomass concentration

1 ml of the sample is taken and subjected to a centrifugation process for 5 minutes at 5000 r.p.m., after which the supernatant is collected on the surface and washed twice with water and then placed in a drying oven at 105°C, the drying continues until the weight is stable [36].

2.7.3 Determine the fermentation power of the obtained yeast

6.75 g of the sugar-phosphate mixture was mixed with 75 ml of calcium sulfate solution in the beaker. Then add 0.893 g of dry baker's yeast. Stir well to disperse the yeast. Then the fermentation power was measured using fermentometer (RHEO FERMENTOMETER F4) [37].

2.8 Modeling

In order to fit the experimental data, three kinetic models (Monod, Verhulst, and Tessier) were chosen.

Monod kinetic model is a substrate concentration-dependent, Verhulst kinetic model is an unstructured model that depends on biomass, and Tessier is an unstructured model for a substrate concentration-dependent [32].

The Kinetic parameters (μ_{max}, K_s, and X_m), were determined after obtaining the curve fitting method of each model performed using Excel software (2016 Microsoft Corporation), and the results showed in **Table 2**, [38].

2.9 Profile prediction of biomass and substrate concentration

The integration of the Verhulst model was used (Eq. (3)), in order to predict the experimental profile of biomass of *S. cerevisiae* during time [32].

$$X = \left(x_0{}^{*}\ {exp^{\mu}_{m}}^{*t}\right) / \left(1 - \left(x_0/x_m\right)^{*}\left(1 - {exp^{\mu}_{m}}^{*t}\right)\right) \tag{3}$$

Kinetic Models	Equations	Linearized form	Symbols
Monod model	$\mu = \mu_{max}{}^{*}(s/(s + k_s))$	$(1/\mu) = (k_s/\mu_{max}) + (1/s) + (1/\mu_{max})$	μ: is the specific growth rate (h^{-1}). μ_{max}: is the maximum specific growth rate (h^{-1}). K_S: is the half-saturation constant (g/L). S: is the concentration in limiting substrate (g/L). X: is the biomass concentration (g/L). X_m: is the Maximum biomass concentration (g/L).
Verhulst model	$\mu = \mu_{max}{}^{*}(1-x/x_m)$	$\mu = \mu_{max}-(\mu_{max}/x_m)^{*}x$	
Tessier model	$\mu = \mu_{max}{}^{*}(1-exp^{-k_s{}^{*}s})$	$ln(\mu) = (1/k_s)^{*}s + ln(\mu_{max})$	

Table 2.
Unstructured kinetic models to determine the kinetic parameters. [32].

The substrate model (Leudeking Piret) as described below (Eq. (4)) was also applied to predict an experimental profile for total reducing sugars consumption by *S. cerevisiae* during the time fermentation.

$$-ds/dt = p^*\,(dx/dt) + q^*x \tag{4}$$

Where ($p = 1/y_{x/s}$) and q is a maintenance coefficient ($q = \mu_{max}/y_{x/x0}$.) Eq. (4) is rearranged as follows:

$$-\mathrm{ds} = \mathrm{p}^*\,\mathrm{dx} + \mathrm{q}\int \mathrm{x}_{(t)}{}^*\,\mathrm{dt} \tag{5}$$

Substituting Eq. (3) in Eq. (5) and integrating with initial conditions ($S = S_0$; $t = 0$) give the following Equation:

$$\begin{aligned} S = s_0 - p\,x_0\,\left\{ {}_{\exp}{}^{\mu}{}_{m}{}^{*t}/1 - (x_0/x_m) * \left(1 - exp^{\mu_m * t}\right)\right\} \\ -q * (x_m/\mu_m) * \ln\,(1 - x_0/x_m) * (1 - exp^{\mu_m * t} \end{aligned} \tag{6}$$

3. Results and discussion

The improvement of dry yeast biomass production was studied by determining the optimum values of the following factors (temperature, initial pH, concentration of sugars in grape juice, the ratio of carbon to nitrogen, and initial concentration of yeasts) that have their influence on the production process using the central composite experimental design, and the central composite design for biomass production in **Table 3**.

Ammonium sulfate and urea were added as a source of nitrogen in a ratio of (50–50) %, taking into account the achievement of the specified ratio between carbon and nitrogen for each experiment, and the agitation speed used during fermentation was 200 r.p.m.

Using the results obtained in diverse experiments, the correlation gives the influence of temperature (x_1), initial pH (x_2), total sugar concentration (x_3), the ratio of carbon to nitrogen (x_4), and initial concentration of yeasts (x_5) on the response. This correlation is obtained by Minitab 19 software and expressed by the following second-order polynomial (Eq. (7)).

$$\begin{aligned} Y = &-261.1 + 8.96\,T + 16.10\,pH + 0.353\,C + 6.55\,C/N + 49.8\,X \\ &- 0.1527\,T^*T - 1.769\,pH^*\,pH - 0.001414\,C^*C \\ &- 0.3025\,C/N^*C/N - 9.30\,X^*X + 0.0316\,T^*\,pH + 0.00096\,T^*C \\ &+ 0.0206\,T^*C/N - 0.117\,T^*X + 0.00414\,pH^*C - 0.0390\,pH^*C/N \\ &- 0.165\,pH^*X + 0.00163\,C^*C/N + 0.0096\,C^*X - 0.016\,C/N^*X \end{aligned} \tag{7}$$

Table 4 shows the coefficient regression corresponding with t and p-values for all the linear and the analysis of variance (ANOVA), quadratic, and interaction effects of parameters tested. A positive sign in the t-value indicates a synergistic effect, while a negative sign represents an antagonistic effect of the parameters on the biomass concentration [39].

Run	Actual Values					(Yi): Biomass (g/L)	
	Temperature (°C)	Initial pH	Concentration of sugars (g/L)	The ratio of carbon to nitrogen	Initial concentration of yeasts (g/L)	experimental Value	Predicted Value
01	35.00	6.000	200.0	8.000	2.000	22.41	23.0429
02	25.00	6.000	200.0	15.000	2.000	20.81	23.1708
03	25.00	3.000	100.0	15.000	2.000	20.01	19.6875
04	25.00	6.000	200.0	15.000	3.000	21.54	24.1742
05	25.00	3.000	200.0	8.000	2.000	19.18	18.5480
06	35.00	6.000	200.0	15.000	3.000	23.02	25.4896
07	25.00	6.000	200.0	8.000	2.000	20.02	21.9975
08	25.00	3.000	100.0	15.000	3.000	19.91	20.2285
09	30.00	4.500	150.0	11.500	2.500	40.45	38.5060
10	30.00	4.500	150.0	11.500	2.500	40.45	38.5060
11	35.00	3.000	200.0	8.000	3.000	18.91	19.0818
12	25.00	3.000	200.0	15.000	2.000	19.71	20.5400
13	35.00	3.000	100.0	15.000	2.000	18.84	20.2692
14	35.00	3.000	100.0	8.000	3.000	17.73	17.4531
15	30.00	4.500	150.0	11.500	2.500	40.45	38.5060
16	35.00	6.000	100.0	15.000	3.000	18.76	21.4784
17	35.00	3.000	200.0	8.000	2.000	17.79	18.6446
18	25.00	6.000	100.0	15.000	2.000	20.11	21.0771
19	35.00	6.000	100.0	8.000	2.000	20.23	21.1317
20	25.00	6.000	100.0	15.000	3.000	20.81	21.1218
21	35.00	6.000	100.0	8.000	3.000	20.91	20.1139
22	35.00	3.000	200.0	15.000	2.000	22.07	22.0803
23	25.00	6.000	100.0	8.000	2.000	21.06	21.0451
24	25.00	6.000	200.0	8.000	3.000	21.92	23.1122
25	35.00	6.000	200.0	15.000	2.000	23.07	25.6599
26	25.00	6.000	100.0	8.000	3.000	20.61	21.2010
27	35.00	3.000	200.0	15.000	3.000	21.41	22.4063
28	25.00	3.000	200.0	15.000	3.000	20.67	22.0397
29	30.00	4.500	150.0	11.500	2.500	40.45	38.5060
30	30.00	4.500	150.0	11.500	2.500	40.45	38.5060
31	30.00	4.500	150.0	11.500	2.500	40.45	38.5060
32	35.00	6.000	100.0	15.000	2.000	23.00	22.6074
33	35.00	3.000	100.0	8.000	2.000	21.11	17.9747
34	35.00	3.000	100.0	15.000	3.000	21.93	19.6364

Run	Actual Values					(Yi): Biomass (g/L)	
	Temperature (°C)	Initial pH	Concentration of sugars (g/L)	The ratio of carbon to nitrogen	Initial concentration of yeasts (g/L)	experimental Value	Predicted Value
35	25.00	3.000	200.0	8.000	3.000	20.27	20.1589
36	30.00	4.500	150.0	11.500	2.500	40.45	38.5060
37	30.00	4.500	150.0	11.500	2.500	40.45	38.5060
38	35.00	6.000	200.0	8.000	3.000	22.03	22.9839
39	25.00	3.000	100.0	8.000	2.000	20.17	18.8368
40	25.00	3.000	100.0	8.000	3.000	20.91	19.4890
41	30.00	4.500	150.0	11.500	2.500	40.45	43.9227
42	41.83	4.500	150.0	11.500	2.500	23.81	22.8190
43	30.00	4.500	150.0	11.500	3.683	33.02	31.1860
44	30.00	4.500	268.3	11.500	2.500	32.17	26.3327
45	30.00	4.500	150.0	11.500	1.317	31.56	30.6159
46	18.17	4.500	150.0	11.500	2.500	24.07	22.2828
47	30.00	4.500	150.0	11.500	2.500	40.45	43.9227
48	30.00	4.500	150.0	11.500	2.500	40.45	43.9227
49	30.00	8.049	150.0	11.500	2.500	30.94	24.7659
50	30.00	0.951	150.0	11.500	2.500	15.11	18.5059
51	30.00	4.500	150.0	11.500	2.500	40.45	43.9227
52	30.00	4.500	31.7	11.500	2.500	18.87	21.9291
53	30.00	4.500	150.0	3.219	2.500	19.11	21.1956
54	30.00	4.500	150.0	19.781	2.500	30.03	25.1663

Table 3.
The central composite design for biomass production.

3.1 Model summary

S: represents the standard deviation of the distance between the data values and the fitted values, the lower the value of S, the better the model describes the response. R-sq (R^2): is the percentage of variation in the response that is explained by the model, the higher the R^2 value, the better the model fits your data. R^2 is always between 0% and 100%. R-sq (adj): Adjusted R2 is the percentage of the variation in the response that is explained by the model. R-sq (pred): Predicted R^2 is calculated with a formula that is equivalent to systematically removing each observation from the data set, estimating the regression equation, and determining how well the model predicts the removed observation. The value of the predicted R^2 ranges between 0% and 100%. By referring to the values obtained in the current study for these parameters, we find that the current study model is acceptable.

The examination of **Table 4** shows that all coefficient regression of the quadratic terms are statistically significant $p \leq 0.05$ and negatively affect the biomass production (**Figure 2**). In contrast, the interaction terms (T, C/N, X, T* pH, T*C, T*C/N, T*X, pH *C, pH *C/N, pH *X, C*C/N, C*X, C/N*X) are statistically not significant $p > 0.05$, and the interaction terms (pH, C, T*T, pH * pH, C*C, C/N*C/N, X*X) are significant with p ˂0.05 and have a synergistic effect on the response.

It is known that the F-value with a low probability p-value indicates a high significance of the regression model [40].

Looking at the analysis of variance (ANOVA), the study shows that the model is important as the F-value had a low probability p-value (p = 0.000), and the resulting value of R^2 was equal to 92.9% and this indicates that only 7.1% of the variance is not explained by the model and therefore there is a good agreement between the model and the experimental data [41]. **Figure 3** shows the fit between the model and experimental data of cell growth.

By reviewing previous studies, Bennamoun et al. [42] used response surface methodology in order to improve and optimization of the medium components,

Term	DF	Adj SS	Adj MS	Coef	SE Coef	T-Value	P-Value	VIF	P-Value
T	1	0.55	0.555	0.113	0.447	0.25	0.802	1.00	0.802
pH	1	75.60	75.595	1.323	0.447	2.96	0.006	1.00	0.006
C	1	37.41	37.408	0.931	0.447	2.08	0.045	1.00	0.045
C/N	1	30.42	30.415	0.839	0.447	1.88	0.070	1.00	0.070
X	1	0.63	0.627	0.120	0.447	0.27	0.789	1.00	0.789
T*T	1	869.04	869.040	−3.818	0.381	−10.03	0.000	1.01	0.000
pH * pH	1	945.05	945.046	−3.981	0.381	−10.46	0.000	1.01	0.000
C*C	1	745.29	745.295	−3.536	0.381	−9.29	0.000	1.01	0.000
C/N*C/N	1	818.56	818.560	−3.705	0.381	−9.74	0.000	1.01	0.000
X*X	1	322.63	322.625	−2.326	0.381	−6.11	0,000	1.01	0.000
T* pH	1	1.80	1.800	0.237	0.519	0.46	0.651	1.00	0.651
T*C	1	1.84	1.838	0.240	0.519	0.46	0.648	1.00	0.648
T*C/N	1	4.17	4.169	0.361	0.519	0.69	0.492	1.00	0.492
T*X	1	2.76	2.755	−0.293	0.519	−0.56	0.576	1.00	0.576
pH *C	1	3.08	3.081	0.310	0.519	0.60	0.554	1.00	0.554
pH *C/N	1	1.34	1.341	−0.205	0.519	−0.39	0.696	1.00	0.696
pH *X	1	0.49	0.493	−0.124	0.519	−0.24	0.813	1.00	0.813
C*C/N	1	2.60	2.605	0.285	0.519	0.55	0.587	1.00	0.587
C*X	1	1.84	1.838	0.240	0.519	0.46	0.648	1.00	0.648
C/N*X	1	0.02	0.025	−0.028	0.519	−0.05	0.958	1.00	0.958

S	R-sq	R-sq(adj)	R-sq(pred)
2.93825	92.85%	88.16%	69.22%

Table 4.
Estimated regression coefficients of t and p and analysis of variance (ANOVA).

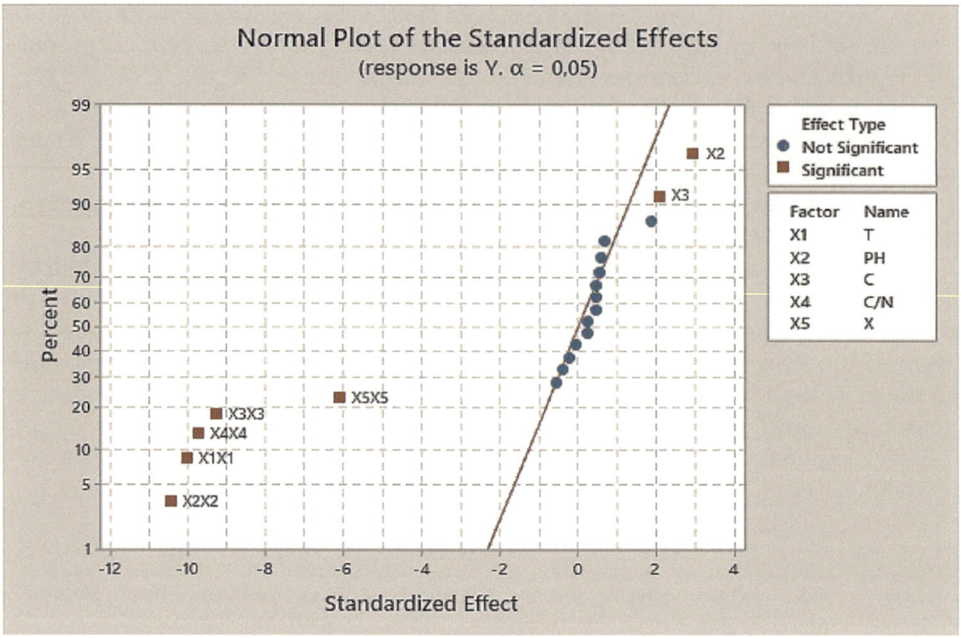

Figure 2.
Variable effect signification on biomass production.

which enhance the polygalacturonase activity of the strain Aureobasidium pullulans, and they got good results (a very low p-value (0.001) and a high coefficient of determination (R^2 = 0.9421), the results confirm the importance and success of using this method.

A previous study by Boudjema, Fazouane-Naimi, and HellaL [27] showed the success of using the experimental design method in the study of the production of *Saccharomyces cerevisiae* DIV13-Z087°CVS using sweet cheese serum, as it confirmed a high significance of the regression model, and the results showed a good agreement with experimental data (a low probability p-value ≤0.000 and a good correlation coefficient (R^2 = 0.914%).

The optimization of the response Yi (Biomass production) and the prediction of the optimum levels of (temperature, initial pH, concentration of sugars in grape juice, the ratio of carbon to nitrogen, and initial concentration of yeasts) were obtained. This optimization resulted in surface plots (**Figure 4**), the figure shows that there is an optimum, located at the center of the field of study.

In addition, the use of the Minitab optimizer will give exact values of the optimum operating conditions of the process **Figure 5**.

Figure 5 shows the maximum biomass production by *Saccharomyces cerevisiae* (41.444 g/L) corresponding to values of temperature (30.11°C), pH (4.75), sugar concentration (158.36 g/L), the ratio of carbon to nitrogen (11.9), initial yeast concentration (2.5 g/L). The amount of urea was 6.65 g/L and the amount of ammonium sulfate used was 6.65 g/L, so that the concentration of added urea and ammonium sulfate was (50–50)% and the required C/N ratio was achieved, and the stirring speed was equal to 200 r.p.m. during the fermentation process. Jiménez-Islas et al. [36] obtained the highest cell concentration of *S. cerevisiae* ATCC 9763 (7.9 g/L) after 26 h

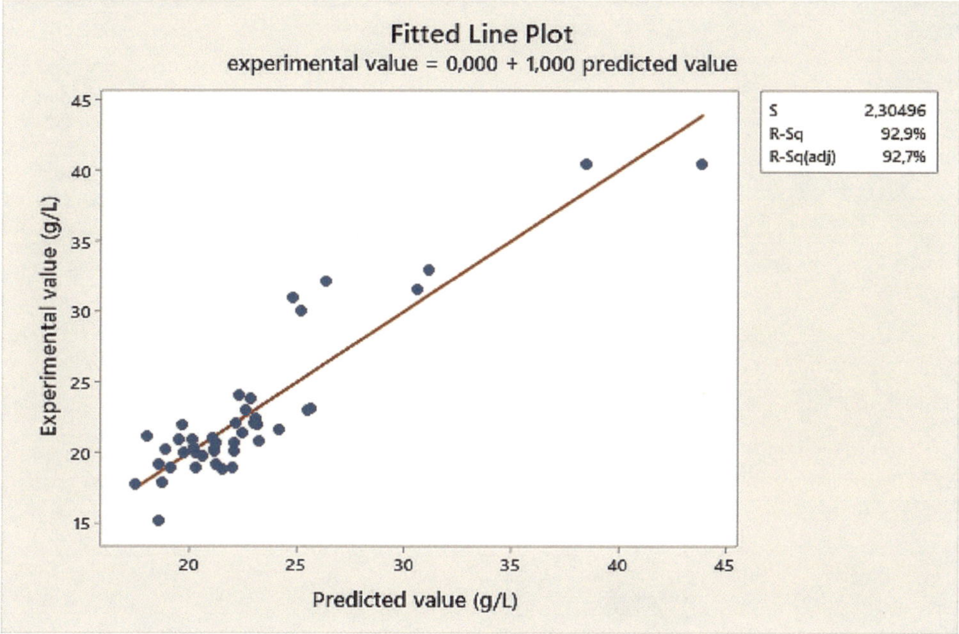

Figure 3.
The fit between the model and experimental data of cell growth.

when the strain grew at 30°C and pH 5.5, so we note that our study gave a good result in achieving the greatest possible production of baker's yeast.

The validation of the baker's yeast biomass concentration and total reducing sugar consumption, over time fermentation, at optimized conditions, are presented in **Figure 6**.

At the beginning of the fermentation process, the concentration of the resulting biomass increases and is associated with the consumption of sugar. After 12 hours of the fermentation process, the sugar concentration has reached a very low level, and this is associated with a decrease in yeast production.

The same results were obtained by Ali et al. [32] where they study the optimization of Baker's Yeast production on Date extract using Response Surface Methodology (RSM), and the resulting yeast was equal to 40 g/L.

The measured fermentation power of the yeast obtained in this study from grape juice was 480 ml, so this is considered to have good fermentation capacity and is suitable for industrial use. The acceptable fermentation strength of yeast is not less than 350 ml according to the COFALEC (2012): General characteristics of dry baker's yeast.

Depending on the Monod model, the curve fitting of cell growth is formed (1/μ versus 1/S) and shown in **Figure 7**. **Figure 8** shows the resulting graph according to the Verhulst model (μ versus X), and in **Figure 9** the graphical curve is formed according to the growth of the Tessier model (μmax and Ks).

The kinetic parameters of growth of *Saccharomyces cerevisiae* using different kinetic models according to the curve fitting method are presented in **Table 5**.

The results obtained from the modeling process appear as follows: the Monod model gave a good value for the parameter R^2 equal to 0.94, which indicates that it is an acceptable model for studying the kinetic performance of a strain *S. cerevisiae*, and

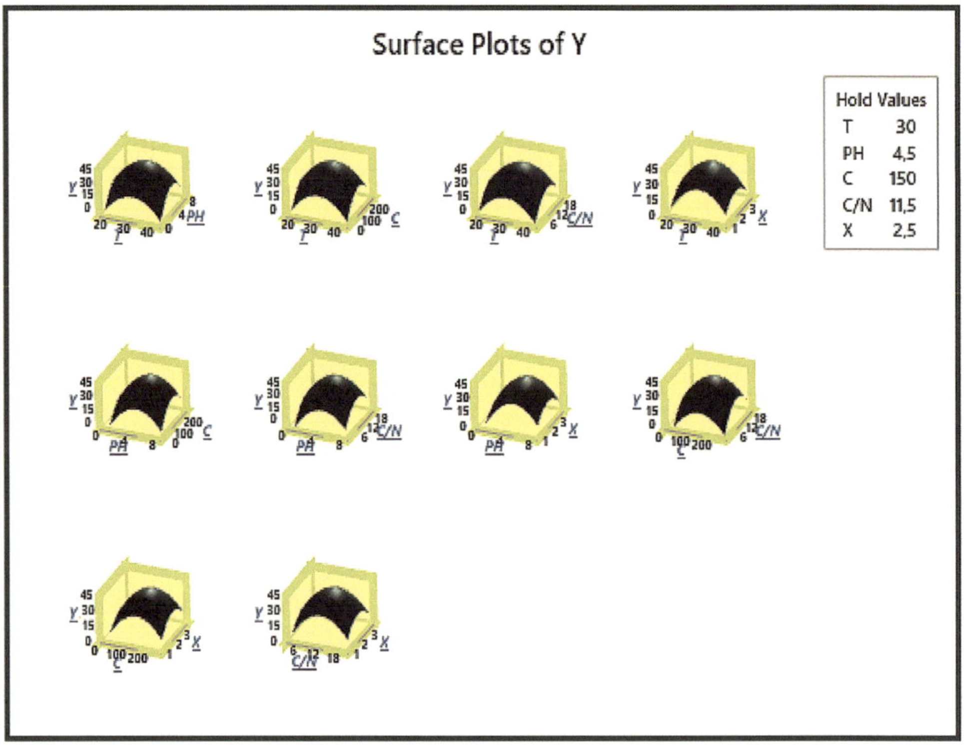

Figure 4.
Surface plot for the effect of different parameters on biomass production.

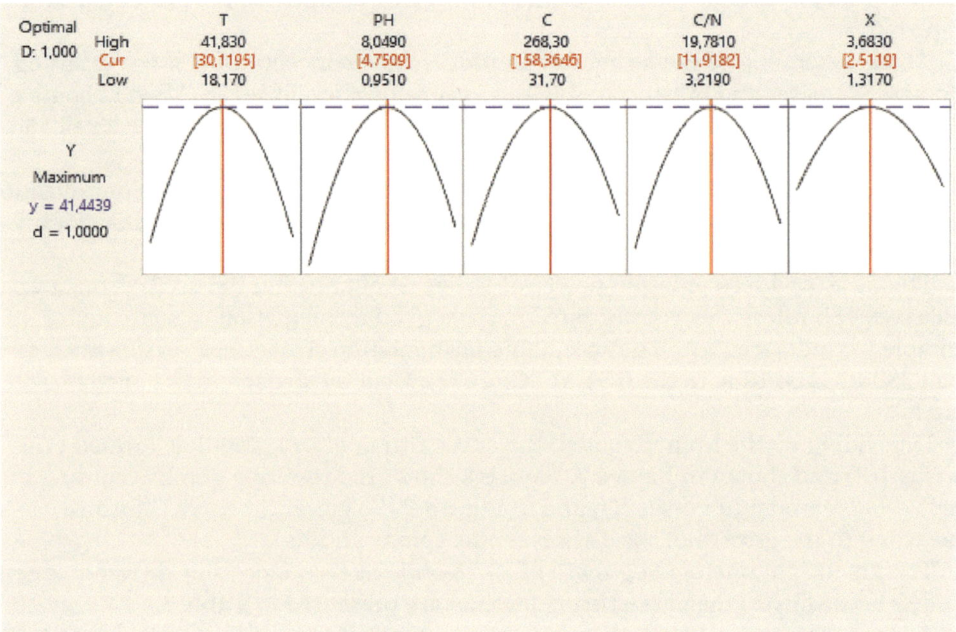

Figure 5.
Values of optimal conditions on biomass production.

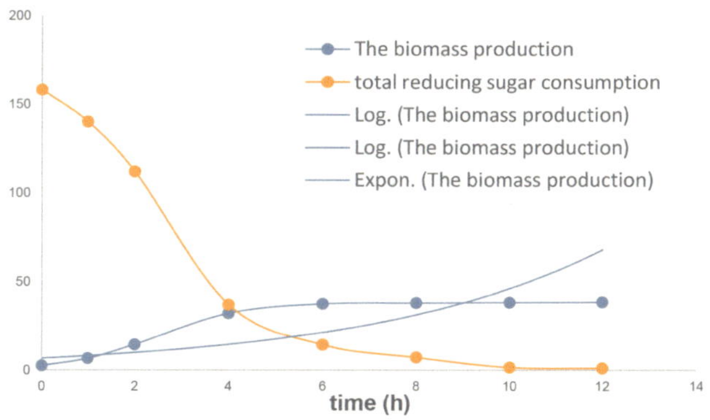

Figure 6.
The biomass production, and total reducing sugar consumption over time at optimized conditions.

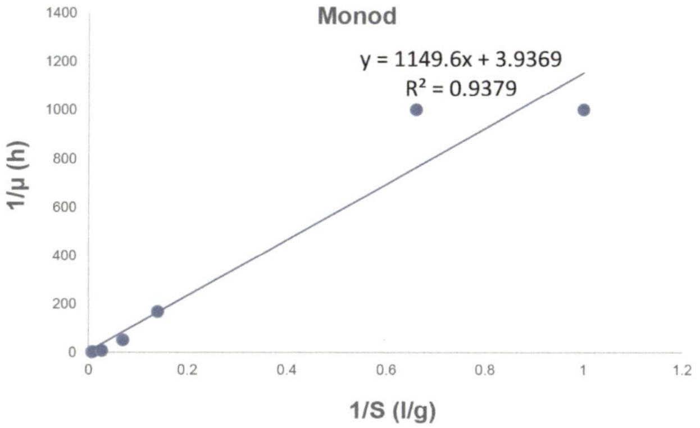

Figure 7.
The line weaver Burk linear plot fitting the experimental data using the Monod kinetic model.

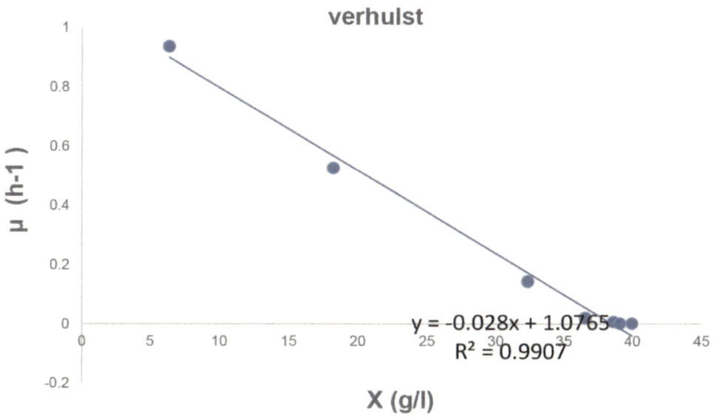

Figure 8.
A plot fitting the experimental data using the Verhulst kinetic model.

the values of each of the maximum specific growth rate (μmax) and is the half-saturation constant (Ks) were evaluated as 0.254 h^{-1} and 291.99 g/L, respectively, which are good values indicating rapid growth of cells Yeast. Tessier's model gave the lowest value for R^2 compared to the Monod and Verhulst models, where it was 0.81. Whereas the Verhulst model gave the highest value for the parameter R^2 which

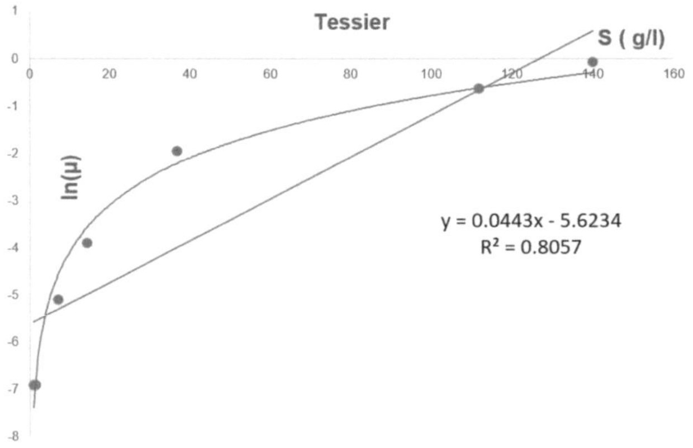

Figure 9.
A plot fitting the experimental data using the Tessier kinetic model.

reached 0.99, also gave a high value for the maximum specified growth rate reached 1.0765 h^{-1}, and the highest possible amount was obtained from the concentration of yeast according to the Tessier model reached 38.26 g/L. As a result, the Verhulst model is the best model for studying and controlling the kinetic behavior of a yeast strain *S. cerevisiae*.

A residual plot is a chart used to assess the quality of a regression fit. Examination of the remaining squares will help determine if the least-squares assumptions are ever met. When these assumptions are met, least squares regression typically yields an inaccurate estimation coefficient with minimal variance. The 4-in-1 residual plot displays four residual plots in a graph window. This configuration can be useful for comparing plans to determine if the Verhulst model meets the criteria for analysis. The remaining sections of the figure are:

- Histogram - indicates if the data is biased or outliers are contained in the data.
- Normal probability plot - indicates whether the data conforms to the normal distribution, whether other changes are affecting the response, or whether the content of the data.
- Residuals versus fitted values - indicates if the difference is continuous, if there is a nonlinear relationship, or if outliers are present in the data.
- -Residuals versus order of the data - indicates whether there is an impact on data due to the time or order of data collection.

Kinetic models	**Parameters of estimation**			
	R^2	K_S (g/L)	μ_{max} (h^{-1})	X_m
Monod	0.94	291.99	0.254	—
Verhulst	0.99	—	1.0765	38.26
Tessier	0.81	22.7	0.0036	—

Table 5.
Kinetic parameters of Saccharomyces cerevisiae *growth and substrate utilization using unstructured models.*

Minitab provides the following residual plots in **Figure 10**.

Examination of the remains indicates that there is nothing to complain about. The normal performance of the remaining sections does not seem to have much difference. There is nothing surprising here and it seems acceptable.

The kinematic models describe the growth rate of microorganisms based on biomass and substrate concentration and are useful because they help engineers design and control biological processes, including the Verhulst model which describes the experimental data obtained on the growth rate of yeast cells, where it describes the logarithmic growth of cells and shows that the first six hours of fermentation were during the initial cell growth phase, then the logarithmic growth phase began, which is characterized by a doubling of the number of yeast cells and an increase in the growth rate.

A profile of biomass and total reducing sugar concentration during fermentation time is compared to the values predicted by the equations model obtained in **Figures 11** and **12**.

During the fermentation, values of biomass between predicted and experimental data were approximately the same. And for total reducing sugar concentration, the values obtained by the Leudeking Piret model were identical to the predicted values, where the values ($p = 1/y_{x/s}$, $q = \mu/y_{x/x0}$) were 3.81 g/g and 0.065 1/h, respectively.

On the basis of these results, good correlation coefficients showed that the proposed Verhulst model and the Luedeking Piret model were adequate to explain the development of the biomass production process in grape juice.

This study confirmed that the Logistic equation for the growth and the Leudeking Piret kinetic model for substrate utilization were able to fit the experimental data, and

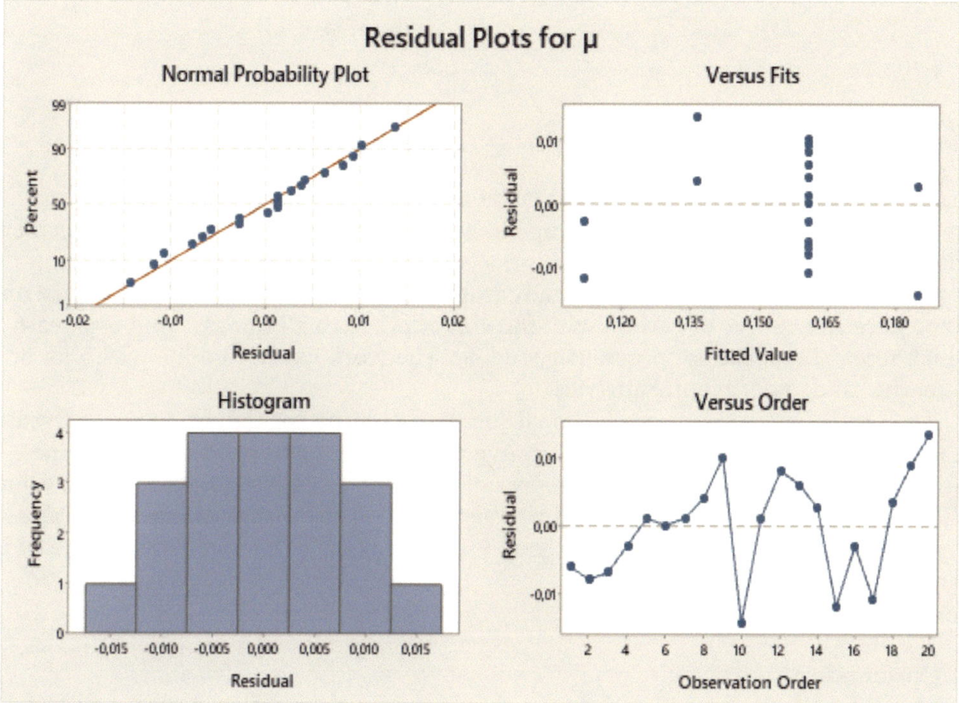

Figure 10.
Residual plots for response.

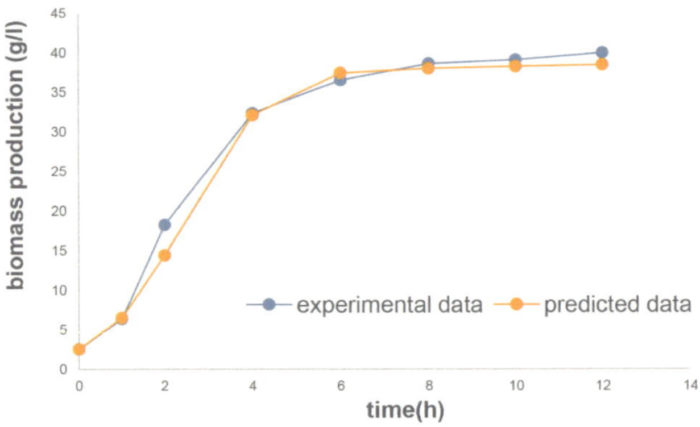

Figure 11.
The comparison between predicted and experimental data for biomass production of baker's yeast.

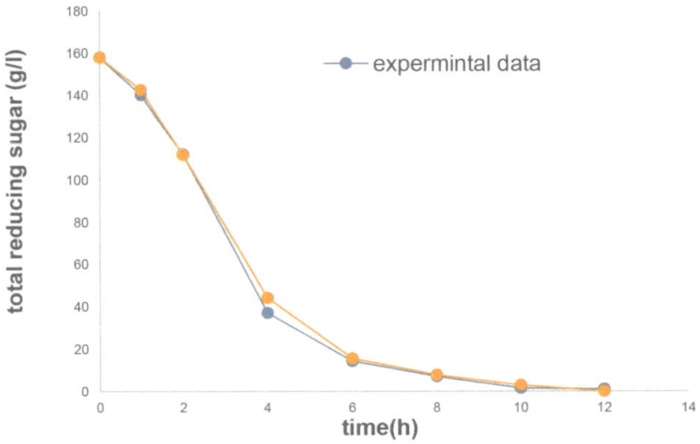

Figure 12.
The comparison between predicted and experimental data for total reducing sugar consumption.

the same result was obtained by Kara Ali et al. [43] Where they used the logistic empirical kinetic model and Leudeking Piret model and they obtained good agreement with the experimental data.

Finally, what distinguishes this study from previous studies is the dependence on grape juice as a source of carbon with the aim of producing biomass from dry yeast, which researchers had not previously studied. The work has been done with a lot of numerical and experimental analysis.

This study will present an additional successful option for the production of yeast that commonly uses molasses. The improvement of the initial conditions of fermentation also contributed to the highest possible yield of yeast and good economic value. The fermentation power of the yeast was also good, so this study can be practically applied with the aim of producing a good mass of baker's yeast and using this yeast in various industrial and food fields.

4. Conclusion

The central composite design (CCD) proposed in this study seems pertinent to describe the optimum biomass production of *Saccharomyces cerevisiae*. A second-order polynomial model was developed to evaluate the quantitative effects of temperature, initial pH, and concentration of sugars in grape juice, the ratio of carbon to nitrogen, initial concentration of yeasts in order to discover the optimum conditions for the biomass production from grape juice. According to the experimental results, a maximum biomass concentration of (41.444 g/L) corresponding to values of temperature (30.11°C), pH (4.75), sugar concentration (158.36 g/L), the ratio of carbon to nitrogen (11.9), initial concentration of yeasts (2.5 g/L), the amount of urea was 6.65 g/L and the amount of ammonium sulfate used was 6.65 g/L, so that the concentration of added urea and ammonium sulfate was (50–50)%, and the used agitation speed was equal to 200 r.p.m. during the fermentation process. The fermenter power of the obtained yeast was 470 ml. In addition, among three unstructured kinetic models, the Verhulst model was the most suitable model to signify the baker's yeast production on grape juice medium.

Acknowledgements

The authors are thankful to everyone supported our work, and to every who collaboration and assistance to carry out this study.

Conflicts of interest

The authors declare no conflict of interest.

Author details

Sawsan Mahmood[1]*, Ali Ali[1], Ayhem Darwesh[1] and Wissam Zam[2]

1 Tartous University, Faculty of Technical Engineering, Tartous, Syria

2 AL- Wadi University, Faculty of Pharmacy, Homs, Syria

*Address all correspondence to: sawsanmahmood480@gmail.com

IntechOpen

© 2022 The Author(s). Licensee IntechOpen. This chapter is distributed under the terms of the Creative Commons Attribution License (http://creativecommons.org/licenses/by/3.0), which permits unrestricted use, distribution, and reproduction in any medium, provided the original work is properly cited. CC BY

Reference

[1] Michela P, Massimo B, Arianna G, Debora G, Cristina S, Mike Q, et al. Characterization and selection of functional yeast strains during sourdough fermentation of different cereal wholegrain flours. Scientific Reports. 2020;**10**:12856. DOI: 10.1038/s41598-020-69774-6

[2] Tamang JP, Watanabe K, Holzapfel WH. Diversity of microorganisms in global fermented foods and beverages. Frontiers in Microbiology. 2016;7:377. DOI: 10.3389/fmicb .2016.00377.28pp

[3] Gelinas P, McKinnon C. Fermentation and microbiological processes in cereal foods. In: Kulp K, Ponte JG, editors. 2nd ed. New York: Marcel Dekker Inc; 2000. pp. 741-754

[4] Phaff HJ. In: Labeda DP, editor. Isolation of Yeasts from Natural Sources in Isolation of Biotechnological Organisms from Nature. New York: McGraw-Hill; 1990

[5] Goffeau A, Barrell BG, Bussey H, Davis RW, Dujon B, Feldmann H, et al. Life with 6000 genes. Sci. 1996;**74**: 563-567

[6] Chris TH, James LS, David SR. Diverse yeasts for diverse fermented beverages and foods. United Nations: Elsevier Ltd.; 2018;**49**:199-206. DOI: 10.1016/j.copbio.2018.10.004

[7] Sivasakthivelan P, Saranraj P, Sivasakthi S. Production of bioethanol by Zymomonas mobilis and *Saccharomyces cerevisiae* using sunflower head wastes–A comparative study. International Journal of Microbiology Research. 2014;**5**(3): 208-216

[8] Prem Kumar D, Jayanthi M, Saranraj P, Kavi Karunya S. Effect of calcium propionate on the inhibition of fungal growth in bakery products. Indo-Asian Journal of Multidiscipline Research. 2015;**1**(3):273-279

[9] Beudeker FR, Van Dam HW, Van der Plaat JB, Vellenga K. Developments in baker's yeast production. In: Verachtert H, editor. Yeast Biotechnology and Biocatalysis. New York: Marcel Dekker, Inc.; 1990

[10] Saranraj P, Sivasakthivelan P, Suganthi K. Baker's yeast: Historical development, genetic characteristics, biochemistry, fermentation and downstream processing. Journal of Academia and Industrial Research (JAIR). 2017:158-164. ISSN: 2278-5213

[11] Lisicar J, Sedaghati M, Barbe S. Looking at baker's yeast fermentation through new glasses: The neglected potential of vinasse for biotechnological applications. In: 31st Yeast Conference, Leuven (Belgium), VH Berlin

[12] Lisičar J, Millenautzki T, Scheper T, Barbe S. New trends in industrial baker's yeast fermentation: Recovery of key biomolecules and low-grade heat conversion. Journal of Biotechnology. 2018. p. 23. DOI: 10.1016/j.jbiotec.2018.06.052

[13] Boyacioglu1 H, Ertunc S, Hapoglu1 H. Modelling of Baker's yeast production. International Journal of Secondary Metabolite. 2015. p.15. http://www.ijate.net/index.php/ijsm

[14] Fadel M, Yousif ES, Abdelfattah A, Ola SS, Sarra E. Approach for highly active Baker's yeast product from distilled yeast biomass. Current Science

International. 2020:321-334. DOI: 10.36632/csi/2020.9.2.27

[15] Reed G, Nagodawithana TW. Yeast Technology. 2nd ed. New York: Van Nostrand; 2011

[16] Arshadm K, Khalil R, Rajokam I. Optimization of process variables for minimization of byproduct formation during fermentation of blackstrap molasses to ethanol at industrial scale. Letters in Applied Microbiology. 2018. 32 p. 475410414

[17] Kopsahelis N, Nisiotou A, Kourkoutas Y, Panas P, Nychas GJ, Kanellaki M. Molecular characterization and molasses fermentation performance of a wild yeast strain operating in an extremely wide temperature range. Bioresource Technology. 2019: 1002048544862

[18] Xandé X, Archimède H, Gourdine JL, Anais C, Renaudeau D. Effects of the level of sugarcane molasses on growth and carcass performance of Caribbean growing pigs reared under a ground sugarcane stalks feeding system. Tropical Animal Health and Production. 2020:4211320

[19] Makhoul G, Mahfoud H, Daoub R. Morphological characterization of some grape types in the sheikh Badr region of Tartous governorate. Tishreen University Journal for Research and Scientific Studies - Biological Sciences Series. 2018;**04**(6):8402

[20] Alexeeva YV, Ivanova EP, Bakunina IY, Zvaygintseva TN, Mikhailov VV. Optimization of glycosidases production by Pseudoalteromonasissachenkonii. KMM 3549T. Letters in Applied Microbiology. 2002;**35**:343-346

[21] Patidar P, Agrawal D, Banerjee T, Patil S. Chitinase production by Beauveriafelina. RD 101: Optimization of parameters under solid substrate fermentation conditions. World Journal of Microbiology and Biotechnology. 2005;**21**:93-95

[22] Rajendhran J, Krishnakumar V, Gunasekaran P. Optimization of a fermentation medium for the production of penicillin G acylase from bacillus sp. Letters in Applied Microbiology. 2002; **35**:523-527

[23] Sayyad SA, Panda BP, Javed S, Ali M. Optimization of nutrient parameters for lovastatin production by Monascus purpureus MTCC 369 under submerged fermentation using response surface methodology. Applied Microbiology and Biotechnology. 2007; **73**:1054-1058

[24] Kennedy M, Krouse D. Strategies for improving fermentation medium performance: A review. Journal of Industrial Microbiology & Biotechnology. 1999;**23**:456-475

[25] Chakravarti R, Sahai V. Optimization of compaction production in chemically defined production medium by Penicilium citrinum using statistical methods. Process Biochemistry. 2002;**38**:481-486

[26] Kar B, Banerjee R, Bhattachaaryya BC. Optimization of physicochemical parameters for gallic acid production by evolutionary operation-factorial design techniques. Process Biochemistry. 2002;**37**: 1395-1401

[27] Boudjema K, Fazouane-naimi F, Hella L, A. Optimization of the bioethanol production on sweet cheese whey by Saccharomyces cerevisiae DIV13-Z087°CVS using response surface methodology (RSM). Romania

Biotechnology Letters. 2015;**20**: 10814-10825

[28] Montgomery DC. Design and Analysis of Experiments. 5th ed. New York, NY, USA: John Wiley & Sons, Inc.; 2009

[29] Kocher GS, Uppal S. Fermentation variables for the fermentation of glucose and xylose using Saccharomyces cerevisiae Y-2034 and Pachysolan tannophilus Y-2460. Indian Journal of Biotechnology. 2013;**12**:531-536

[30] Naser FN, Abdelrahman Z. Five Major Factors Affecting the Production of baker's Yeast Using Sugar Cane Molasses. Faculty of Agriculture: Cairo University; 2017 https://www.researchgate.net/publication/315838486

[31] Muhammad, A.K.; Muhammad, M. J.; Qurat, A.; Sana, Z.; Kaleem, I. Process optimization for the production of yeast extract from fresh Baker's yeast (Saccharomyces cerevisiae). Department of Biotechnology, Virtual University, Lahore, Pakistan. 2020, (www.preprints.org).

[32] Ali M, Outiti N, Kaki A, Cherfia R, Benhassine S, Benaissa A, et al. Optimization of Baker's yeast production on date extract using response surface methodology (RSM). MDPI. Foods. 2017;**6**:64. DOI: 10.3390/foods6080064

[33] Taleb T, Khalid A, Abederhman K, Bhaskara T. Optimization of bakery yeast production cultivated on musts of dates. Journal of Applied Sciences Research. 2007;**3**(10):964-971

[34] Sokchea H, Thi Hang P, Dinh Phung L, Duc Ngoan L, Thu Hong T, Borin K. Effect of time, C/N ratio and molasses concentration on saccharomyces cerevisiae biomass production. Journal of Veterinary Animal Research. 2018

[35] Dubois M, Gilles KA, Hamilton JK, Rebers PA, Smith F. Colorimetric method for determination of sugars and related substances. Analytical Chemistry. 1956;**28**:350-356

[36] Jiménez-Islas D, Páez-Lerma J, Soto-Cruz NO, Gracida J. Modelling of ethanol production from red beet juice by Saccharomyces cerevisiae under thermal and acid stress conditions. Food Technology and Biotechnology. 2014;**52**: 93-100

[37] COFALEC. General Characteristics of Dry baker's Yeast. 14 rue de Turbigo, 75001 Paris, France; 2012

[38] Juska A. Minimal models of growth and decline of microbial populations. Journal of Theoretical Biology. 2011;**269**: 195-200

[39] LeMan H, Behera SK, Park HS. Optimization of operational parameters for ethanol production from Korean food waste leachate. International Journal of Environment and Science Technology. 2010;7:157-164

[40] Rene ER, Jo MS, Kim SH, Park HS. Statistical analysis of main and interaction effects during the removal of BTEX mixtures in batch conditions using wastewater treatment plant sludge microbes. International journal of Environmental Science and Technology. 2007;**4**:177-182

[41] Annuar MSM, Tan IKP, Ibrahim S, Ramachandran KB. A kinetic model for growth and biosynthesis of medium-chain-length poly-(3-Hydroxyalkanoates) in pseudomonas putida. Brazilian Journal of Chemical Engineering. 2008;**25**:217-228

[42] Bennamoun L, Hiligsmann S, Dakhmouche S, Ait-Kaki A, Labbani FZK, Nouadri T, et al. Production and properties of a thermostable, PH-stable Exo-Polygalacturonase using Aureobasidium pullulans isolated from Saharan soil of Algeria grown on tomato pomace. Food. 2016;**5**:72

[43] Kara Ali M, Hiligsmann S, Outili N, Cherfia R, Kacem Chaouche N. Kinetic models and parameters estimation study of biomass and ethanol production from inulin by Pichiacaribbica (KC977491). African Journal of Biotechnology. 2017; **16**:124-131

Chapter 8

Response Surface Model Applied to Fine Arts: The Case of the Restoration of Paintings

Julio Romero-Noguera, Nuria Pérez-Villares, Fernando Bolívar-Galiano and Rafael Bailón-Moreno

Abstract

Cleaning polychrome paintings and sculptures is an essential task in restoration treatment, since it irreversibly affects the appearance and material structure of such works of art. It is a completely "analogical" process consisting of removing surface dirt, aged varnishes or repainting (paints added to the original) based on the restorer's experience and knowledge, as well as on different internationally accepted criteria for such interventions. In this chapter we are presenting an example of the adaptation of the response surface model to this field, which is complex and difficult to adapt to quantitative parameters and has never before been studied with this approach. Using the MODDE Go® experiment optimization and statistical design software, the effectiveness of cleaning pictorial works of art has been studied using various formulas composed mainly of water and a low-toxicity monoterpene: limonene. The model's statistical validity is demonstrated, as well as its ability to determine the main factors that affect the cleaning by means of different responses (methods) to evaluate its effectiveness: an expert's opinion using visible light and ultraviolet light, the amount of varnish removed using gas chromatography coupled with mass spectrometry, and the effects on color, lightness and gloss. The main influential factors were the concentrations of the two main components of the proposed formulations, water and limonene, which regulate the cleaners' level of hydrophilia and lipophilicity, followed by the types of pigments and type of varnish used, and aging. Using an *in silico* simulation, the proposed model also enables specific compositions to be formulated for different scenarios and cleaning applications that are potentially effective and harmless to the pictorial materials and the restorers' health.

Keywords: cleaning, oil paintings, water, limonene, response surface

1. Introduction

Since the second half of the 20th century, works of art have been restored based on a fundamentally scientific perspective, using a great variety of products and methods for analysis that have enabled the materials to be characterized in detail, and the results of the restoration treatment to be experimentally proven. However, the results obtained have yet to be correlated with empirical models that adequately back them, which has not yet been studied enough [1–8].

The approach in this study uses an innovative model that compares the nature and conditions of the artwork to be cleaned, the composition of the cleaner and the results obtained after cleaning from multiple perspectives. This involves representing the complex phenomenon of cleaning and stripping varnishes over oil paint by using a model of surface responses. This model can be simulated *in silico* to highlight the synergistic and antagonistic relationships among the main factors involved in cleaning of oil paintings: the type of varnish, degree of aging, type of oil pigment and composition of the cleaner. To do so, different responses (methods) have been brought together to evaluate the cleaning's effectiveness: an expert's opinion using visible and ultraviolet light, the amount of varnish removed using gas chromatography coupled with mass spectrometry, and the effects on color, lightness and gloss. The simulation will also allow optimal cleaning products to be developed for specific cleaning treatments.

1.1 Cleaning works of art

Cleaning is one of the fundamental treatments used in restoring paintings and other types of works of art, and also one of the most controversial ones, since it is the one that most affects their appearance. The term refers to three types of tasks [9]:

- Surface cleaning: removal of non-adhering dirt.

- Varnish cleaning: total or partial removal. Resistant, greasy dirt and the oxidation of a varnish can create a layer where the two are intimately related. These layers can be removed together during the cleaning treatment.

- Lifting of repaints added to the original work and which it has been decided to remove, also known as stripping.

From classical antiquity to today, the criteria used in applying cleaning treatments to artworks have changed along with the development of concepts and theories as regards conservation and restoration.

The lack of control in using cleaning substances has led to the complete or partial loss of polychromy in many artworks. The substances used included highly aggressive products such as soap, diluted bleach and ash. Soda, urine, salt, alum, acids, ox gall, milk and egg yolk, for example, were also products commonly used in the 17th and 18th centuries. Gradually, an awareness of the potential aggressiveness of some of these substances for paints emerged [10].

In the 20th century, a great boost was given to the theory and practice of cleaning cultural assets, mainly due to what had been learned from the alterations caused by many of the products over time, and the risk involved in using solvents. The greatest stimulus came from the scientific advances made after the First World War, which provided a wide variety of products with physical and chemical properties that enabled problems to be solved and new techniques developed [11]. The research carried out in the second half of the century then laid the foundations for more scientifically based restoration work, concentrating on the main solvents' solubilization power as regards the materials to be removed [12–15]. There is now a growing awareness of the danger that cleaning can bring about for the artwork's integrity and the restorers' health.

One alternative to solvents are water-based cleaning systems that include surfactants and other additives in complex detergent formulas [16–18]. It is essential to know the composition of the detergents and the surface to be cleaned in order to determine the effectiveness of the detergent, but even so it is difficult to choose the best cleaner in each case, even for highly trained and experienced restorers.

1.2 Strata involved in cleaning paintings: dirt, varnish and painting layer

The main factors that can alter the appearance of an artwork's color over time are the accumulated dirt on the surface, the darkening and yellowing of the varnish and oil, pigment migration, and the effects of visible and ultraviolet light. When the cleaning is carried out for a polychromy, there are three layers that are affected (**Figure 1**): the dirt, the layer of varnish and the underlying pictorial layer (which in our case study is oil paint), which can alter the artwork's visual appearance [19]. We will briefly review their characteristics.

1.2.1 Dirt

The dirt that we may find on the surface of a painting is a difficult concept to define and varies considerably depending on the circumstances. Surface dirt is understood to mean the sediments that are deposited on an artwork's surface in multiple layers and bound by different forces of attraction [20]. This generally includes particles of dust, carbon and other solid materials such as sand, soil, corrosive products and salts. It is responsible for the grayish veil over the pictorial surface and sometimes causes mechanical damage or reactions with the materials within it as its components

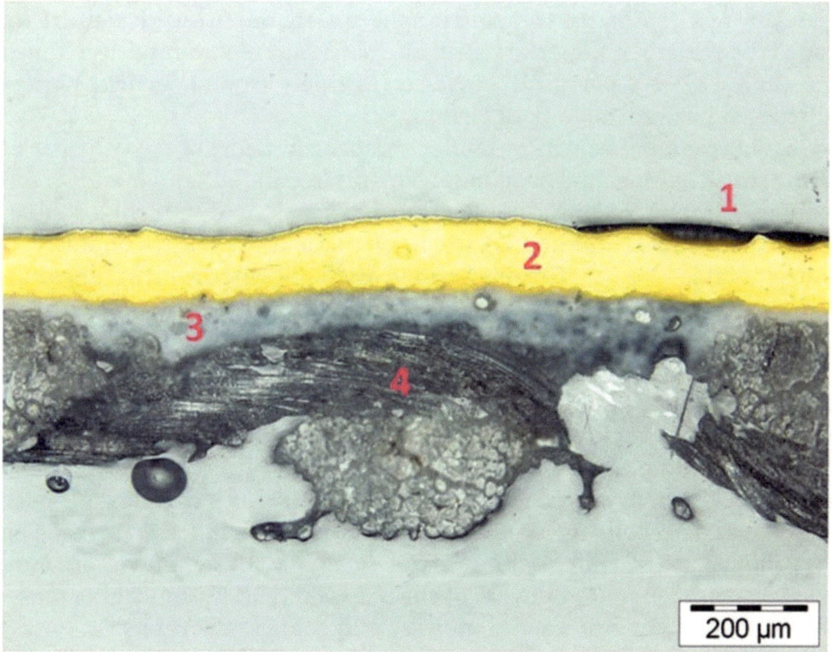

Figure 1.
Strata of a standard sample: [1] varnish layer [2] oil layer, [3] preparation layer, [4] canvas.

absorb some pollutants from the atmosphere. Non-polar surface dirt particles are bound together by weak intermolecular forces, and polar ones by stronger dipolar forces. It is usually sufficient to apply mechanical means and detersive substances in order to remove surface dirt [9].

Surface dirt on works of art is usually associated with fatty deposits made up of a complex mixture of components [21], predominantly natural lipids (triglycerides), which contain unsaturated fatty acids (susceptible to oxidation by air). This type of dirt remains attached to the surface after surface cleaning due to the greater strength of its molecular bonds and interactions [22]. To remove it, it is common to use organic solvents, which can damage the paint layer, both when it is applied and in the long term.

1.2.2 The varnish layer

A varnish is a liquid which, when applied to a solid surface, dries forming a transparent film with varying degrees of gloss, hardness, flexibility, and protection depending on its composition [12]. It is a material of prime importance in the sphere of artistic techniques, which must have an even finish and be transparent, stable and reversible, while preventing efflorescence from developing. Its main purposes in a work of art are for protection and esthetics [23]. The natural varnishes traditionally used in painting are terpenoids, which undergo oxidation processes and other chemical changes that cause them to yellow and lose mechanical and optical properties [24–27]. One of the most frequent painting restoration tasks is to remove aged varnishes by using solvents and replace them with polymeric varnishes, generally acrylics, which are much more stable.

1.2.3 The pictorial layer: oil paint

Oil painting has dominated the artistic sphere since the fifteenth century until today due to the variety of pictorial resources it offers as regards opacity, transparency and chiaroscuro [28]. A layer of oil is made up of finely ground particles of pigment evenly dispersed in a vegetable-based drying oil.

A drying oil is a liquid vehicle or binder composed mainly of triglycerides of fatty acids with 16 or 18 carbon atoms: palmitic, stearic (saturated) and mainly polyunsaturated ones. Among the unsaturated fatty acids, oleic acid (C18, one double bond), linoleic acid (C18, two double bonds) and linolenic acid (C18, three double bonds) are the most notable [29, 30]. The most widely used oils since ancient times have been walnut, poppy and especially flax, since they form transparent films after the drying process, with optimal mechanical and optical properties [31].

The oils dry by oxidation and subsequent polymerization of the triglycerides' unsaturated fatty acids, until they form a relatively hard yet elastic film. After a series of complex chemical reactions involving processes of crosslinking, oxidation of unsaturated acids and the hydrolysis of glyceride bonds, a new substance is formed that is usually called linoxin, with very different physical and chemical properties from the original liquid oil, and which will not return to its initial state by any means [32, 33]. Although the oil film dries out to the touch in weeks, it undergoes new chemical reactions throughout the life of the painting [19]. Natural aging makes the pictorial film less flexible and causes cracking and changes in opacity.

When one intends to clean or remove a varnish from a polychrome surface, it has to be taken into account that the pictorial layer may be altered [34], especially when glazing techniques are applied in the painting's finishing, in which the pictorial medium is a fine mixture of oil, pigments and varnish, and therefore has a composition and polarity that closely resemble the protective varnish that is going to be removed.

Solvents can also give rise to changes in the oil's properties and composition, fostering leaching of components with a low molecular weight such as ketones, alcohols and dicarboxylic acids, like azelaic acid. This process affects the physical properties of the pictorial layer, reducing its volume, increasing its density, and making it brittle and opaque [23].

The type of solvent used for cleaning is decisive. It is generally thought that the greater the polarity of the solvent, the greater the risk of leaching [15], since the oxidation and hydrolysis of the initial triglycerides over time causes changes in the oil paint's chemical structure, making it more polar [35]. The magnitude of the changes also depends on the length of exposure time. When solvents are applied repeatedly or in excessive amounts, they cause surface wear as pigments get washed away with the oily film protecting them. Finally, the nature of the pigment also influences the effect of the solvents on the oil. One well-known example of this is the effect of one of the most significant pigments in art history, lead white, which minimizes the action of solvents even on fairly young oil layers [19].

1.3 Oil painting cleaning treatments

1.3.1 Cleaning methods

Cleaning can be done mechanically or by means of solvents, or else by combining both approaches in mixed treatments. Mechanical cleaning is done with vacuum cleaners, dusters, soft paintbrushes, brushes, compressed air, rubber erasers, lasers or scalpels [36]. It is used for superficial cleaning and as a treatment prior to any intervention in the sphere of restoration, as well as in cases of varnishes, repainting or dirt that is impossible to remove by other means.

Physical and chemical methods involve cleaning with solvents to soften and disperse or solubilize the material to be removed, forming a homogeneous mixture with it. This is finished off with mechanical wiping using a cotton bud or inert media such as cellulose pads or gels that keep the product active for longer. In the sphere of conservation and restoration, these procedures are carried out following internationally accepted cleaning guidelines and standard solubility tests [31].

1.3.2 Solvent properties

There are two different, closely related processes in the action of solvents [37]:

- Initial softening of the substrate by swelling of its molecules' chains.

- Subsequent dispersion or solubilization of the particles that give rise to dissolution.

According to the principles of thermodynamics, each type of substrate must be dissolved by a solvent of similar polarity. It is therefore essential for there to be chemical similarity between the molecules of the solvent and the solute, defined by the predominant intermolecular forces. What is commonly known as "like dissolves like" therefore refers to the fact that a solvent will remove the layer of varnish and/or dirt when it interacts with it with the same type of intermolecular forces as those that hold its own molecules together. Hansen's solubility parameters and visual diagrams such as the Teas triangle are often used to characterize solvents and classify them for their use in restoration [38–40]. Other very important factors must also be considered, such as the penetration capacity, volatility and retention in the artwork, not forgetting the toxicity values for the restorer [12].

2. Response surface model

All the above gives an idea of how enormously complicated it can be to approach the cleaning of artistic paintings from a scientific point of view. There are factors involved that are related to the material, which is chemically very complex and divided into three layers: dirt, varnish and the painting layer. These factors can in turn be subdivided into internal micro-layers with different compositions, as happens when a painting is repainted, in other words, when a new pictorial layer is added to an already finished work. Organic materials also appear, such as binders and varnishes, and also inorganic ones, such as many pigments. We could also distinguish between components that are natural or synthetic, original or added, and polar or non-polar. Metals can even appear if the work includes gilding or silver-plating techniques. Likewise, factors such as aging of the materials to be treated, deterioration agents, or previous restoration treatments are all very important. Lastly, when solvents are being applied, a single product is seldom used, since the habitual values of polarity required in cleaning and stripping varnishes are usually achieved by using solvent mixtures [37]. In restoration practices today, we should also add the frequent use of surfactants, chelating agents or enzymes [14, 21, 37].

Our research aims to analyze the most important factors affecting the effectiveness of cleaning a pictorial work of art so as to be able to put forward effective cleaning methods with few adverse effects. Due to the number of variables present, we used the MODDE Go® (Umetrics) software for statistical design of experiments and optimization, run on a PC with a 64-bit Windows 10 operating system. An effort has been made to include the utmost number of factors and reduce the number of experiments to a minimum, while being as representative as possible of the complex phenomenon that we are attempting to analyze.

As a way of explaining the rationale behind this procedure, think for example of carrying out four experimental points of four different concentrations of five components of a cleaner, plus one point for each of, let us say, five pigments present and two points for each of the factors of aging and the type of varnish. This would mean carrying out at least 20,480 different cleaning tests in the laboratory (45 × 5 × 22 = 20,480). Using statistically designed experiments, the representative sample has been reduced to only 72 cleaning trials. This has meant an enormous saving in time and material resources, which if an attempt had been made to carry out all of the theoretical tests would have made it impossible to actually do them.

The proposed response surface model uses analytical techniques (responses).

of a physical, chemical and visual nature to study the effectiveness of low-toxicity formulations, taking into account the main factors that influence cleaning: the composition of the cleaners, types of pigments and varnishes, and their aging. It also enables in silico simulations in order to develop optimal cleaning products for specific cleaning treatments depending on the characteristics of the pictorial work of art to be restored. Furthermore, in future research, three-dimensional vector models can be developed to analyze: 1) the material and physical–chemical aspects of cleaning, 2) the restoration technique used, and 3) the visual appearance, which can be evaluated using optical methods.

Below, we explain the fundamental points in the proposed design of experiments and some examples of the results obtained. The full technical details of the study can be consulted in Bailón-Moreno et al. [41].

2.1 Preparation of samples

The proposed cleaning methods were tested on reference samples containing the usual layers in an oil painting: support (linen canvas), preparation, paint layer and protective varnish.

The preparation applied over the canvas was composed of animal glue, calcium sulphate ($CaSO_4{\cdot}2H_2O$) and zinc white (ZnO). The oil painting was handmade prepared with stand linseed oil and five different pigments, one for each type of sample: zinc white (ZnO), lead white $(PbCO_3)2Pb(OH)_2$, cadmium yellow (CdS), cadmium red ($3CdS{\cdot}2CdSe$) and cobalt blue ($CoO{\cdot}nSnO_2$). All these products were purchased at Manuel Riesgo, Madrid, Spain) except lead white, wich was produced by ourselves [42].

After a drying period of 3 months, the samples were varnished following two possible procedures: using a traditional terpenoid varnish composed of mastic resin diluted in spirit of turpentine or using an acrylic synthetic varnish by Lefranc & Bourgeois®. In both cases, they were allowed to dry naturally for 12 months (**Figure 2**). Aged terpenoid varnishes such as mastic are affected by

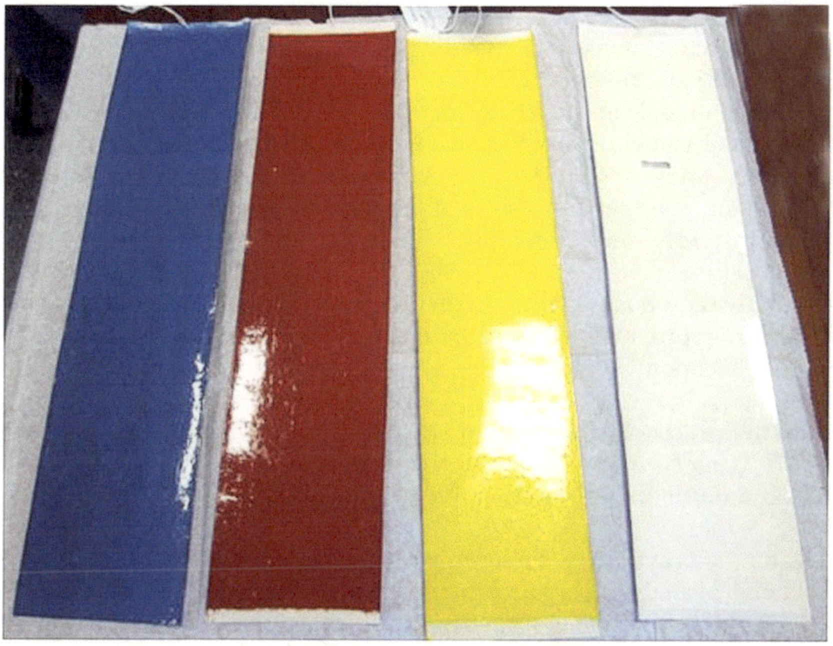

Figure 2.
Reference samples: cobalt blue, cadmium red, cadmium yellow and lead White varnished with mastic.

chemical processes of crosslinking and oxidation that make them more polar than the original ones, and more difficult to remove using solvents in cleaning processes.

To imitate the deterioration of a layer of old paint varnish, some of the samples were subjected to artificial accelerated aging by exposure to ultraviolet light [31]. The rest of the samples were reserved to simulate a recent painting. The varnishes and oil color layers were applied with a micrometric adjustable paint applicator SH-1117/100 (Daesan CMC, South Korea).

2.2 Designing experiments. Software MODDE Go®

The model consists of a set of polynomials (one for each response), which have a constant value, a_0, representing the mean value of the response considered. These terms represent the linear effects of the factors on the responses, $\sum_{i=1}^{8} a_i F_i$, quadratic terms, $\sum_{i=4}^{8} a_{ii} F_{ii}^2$, and finally cross terms, $\sum_{i=1}^{8}\sum_{j=i+1}^{8} a_{ij} F_i F_j$, which represent the synergistic and antagonistic effects between the different factors. S is the response and the values of ai, aj and aij are coefficients that multiply the factors Fi, Fj.

$$\text{Response} = \text{Coeff.} + \{\textit{Lineal eqs}\} + \{\textit{Quadratic eqs}\} + \{\textit{Synerg.\ \&\ antag.eqs}\}$$

$$S = a_0 + \sum_{i=1}^{8} a_i F_i + \sum_{i=4}^{8} a_i F_i^2 + \sum_{i=1}^{8}\sum_{j=i+1}^{8} a_{ij} F_i F_j$$

The model was adjusted with the MODDE Go® software from the company Umetrics using the Partial Least Square (PLS) technique with pseudo-components with non-scaled, non-centred values. Bailón-Moreno et al. [41] show the coefficients associated with each response, S, depending on the model proposed, the coefficient of determination, R2, and the coefficient Q2.

The proposed model considers the cleaning of painted artworks to be a procedure affected by a set of values or variables that is evaluated via a set of responses. The factors can be quantitative or qualitative, depending on whether they can be represented by quantity or not, and they can also be of the process or composition type. The factors chosen are as follows (**Figure 3**).

1. Quantitative composition factors: these define the composition of the cleaner, with different proportions of water, limonene, phenethyl alcohol, Findet ® 1214/N23 and Glucopon® 600.

2. Qualitative process factors: type of varnish (traditional: mastic; or synthetic: acrylic), aging (yes or no), type of pigment in the paint layer (zinc white, lead white, cadmium yellow, cadmium red or cobalt blue).

The cleaning was evaluated via seven possible sets of responses: the physical state, the chemical analysis via gas chromatography/mass spectrometry (GC/MS), cleaning from the point of view of an expert's opinion (observed with visible light and ultraviolet light), and also how the cleaning affects the painting from an optical and colorimetric point of view (color, lightness and gloss) [43, 44]. MODDE Go® (Umetrics), was used to establish a statistical design for experiments in keeping with

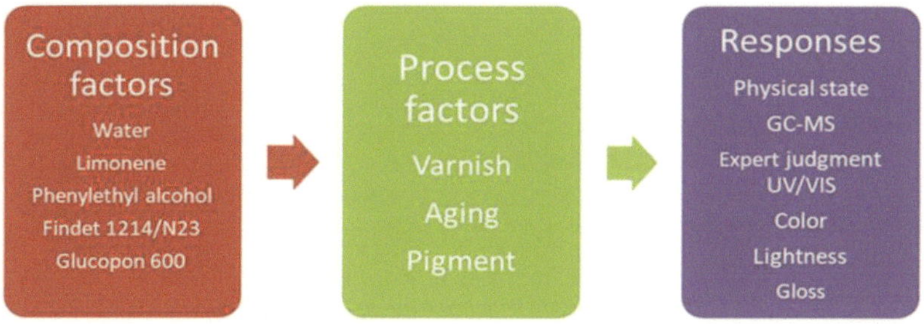

Figure 3.
Response surface model for cleaning oil paintings with composition factors, process factors and the responses studied.

the response surface model put forward. It has 72 statistically representative tests, whose experimental conditions can be consulted in Bailón-Moreno et al. [41]. Every test was performed once.

2.2.1 Quantitative composition factors

These are dependent on the composition of the proposed cleaning mixtures. Several criteria have been used in choosing the products [14, 21, 37].

1. Correct structure of the cleaners. The cleaners must be made up of components that enable stable, effective compositions to be formulated. To do so, the compositions may consist of:

    - One or two main solvents;

    - Optionally a co-solvent;

    - Optionally a surfactant with the possibility of a co-surfactant.

2. A wide variability in the mixtures' polarity. This variability lies in the two main solvents, one polar and one non-polar. Since substances of very different polarities cannot normally be mixed properly by themselves, it is important for there to be a co-solvent of intermediate polarity between the two main solvents, or else one or two surfactants to form an emulsion.

3. Non-toxic and low skin irritation.

4. Easily biodegradable components, in order to avoid environmental problems.

5. They should form compositions that are easy to prepare and use.

6. Industrially affordable and economical components.

7. Components that have already been tested in previous research with good results.

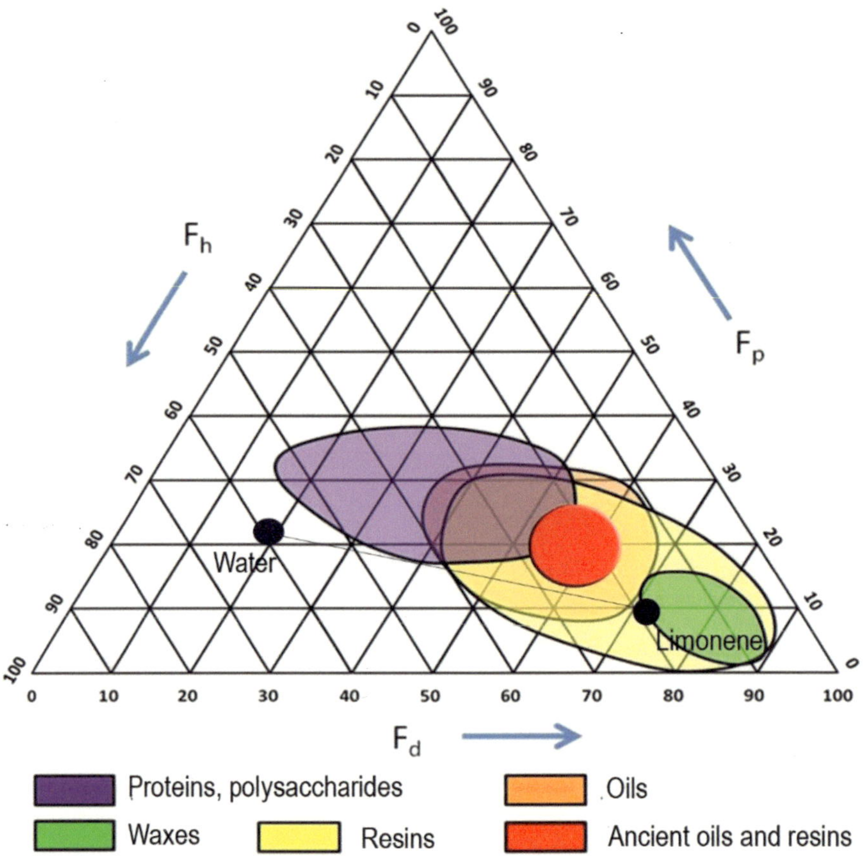

Figure 4.
Masschelein–Kleiner diagram: Solubility of natural film-forming substances and position in the triangle of the two main components of the proposed cleaning formulas: water and limonene. Note how there is partial overlap between resins and oily layers.

In keeping with these general requirements, five substances have been chosen. The proposed cleaning method is based on a mixture of two main components: one clearly polar, water; and the other strongly nonpolar, limonene (1-Methyl-4-(1-methylethenyl)-cyclohexene), a hydrocarbon (monoterpene) devoid of toxicity that is found as the main component in the essential oils of orange, lemon and other aromatic plants. The relative proportion of these components marks the polarity of the mixture and its greater or lesser effectiveness in dissolving each type of material (**Figure 4**).

The formulations have been stabilized by the presence of three products: Findet® 1214/N23 (KAO Chemicals Europe, Barcelona, Spain), comprised of a vegetable-based narrow-range ethoxylate with a C12-C14 fatty chain and 11 moles of ethylene oxide; and Glucopon® 600 (BASF, Barcelona, Spain), a non-ionic surfactant of the alkyl polyglycoside type, specifically a lauryl glucoside with 1.3 moles of glucose.

The cleaning compositions used contain, according to the response surface model, variable amounts of these substances that are statistically representative in all of their possible cleaning formulations. The concentration ranges for each component were Water and limonene: from 0 to 100%, Phenethyl alcohol: from 0 to 5%, Findet® 1214/N23 and Glucopon® 600: from 0 to 10% [41].

Figure 5.
Cleaning process of unaged mastic varnish on cadmium red oil with formulation N44.

2.2.2 Cleaning trials

The cleaning formulations were applied to the samples and allowed to act for 5 minutes. Afterwards, possible residues of the formulations were eliminated by washing with distilled water and subsequently White spirit (Talens). The process was repeated three times on each sample (**Figure 5**).

2.2.3 Responses

The cleaning was evaluated via seven possible sets of responses: the physical state, the chemical analysis via gas chromatography/mass spectrometry (GC/MS), cleaning from the point of view of an expert's opinion (observed with visible light and ultraviolet light), and also how the cleaning affects the painting from an optical and colourimetric point of view (color, lightness and gloss). The complete description of the analytical study can be found in Bailón-Moreno et al. [41] (**Figures 6–8**).

3. Results and discussion

3.1 Validity of the model and most important factors

In order to confirm the validity of the model, the predicted values for the model were compared with the values observed experimentally for each response, achieving very good concordance between the values observed empirically in the 72 experiments actually

Figure 6.
Photographs with UV/(left) and oblique visible light (right) of the standard sample varnished with mastic on cadmium red oil after cleaning with formulation N44.

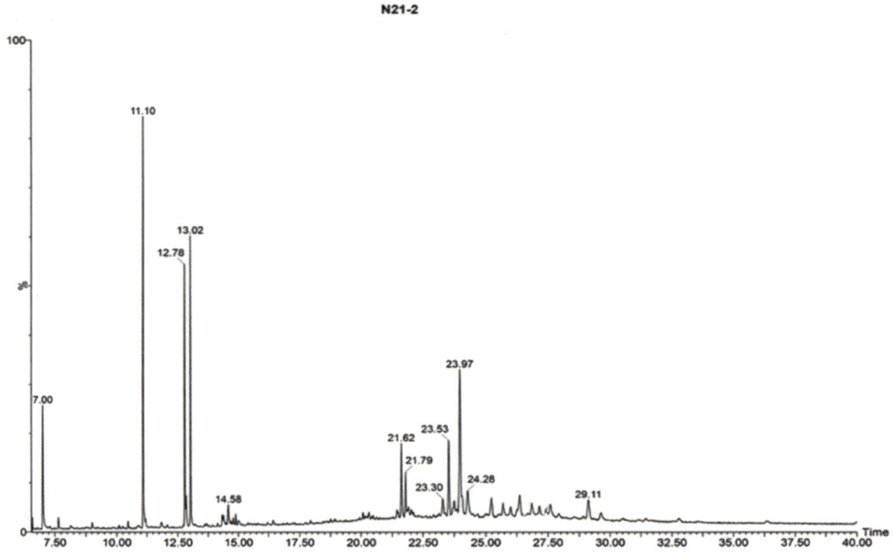

Figure 7.
Chromatogram of a standard sample composed of cadmium yellow oil and aged mastic varnish after cleaning with formulation N21. The peaks corresponding to the fatty acids are observed as main markers of the oil on the left (azelaic acid tR:7, palmitic acid tR:11.1, oleic acid tR:12.78 and stearic acid tR:13.02) and to the triterpenic resin acids as main resin markers on the right (ursonic acid tR:23.53, ursolic acid tR:23.97, moronic acid tR:25.99 and oleanonic acid tR:26.37).

Figure 8.
Gloss measurement of reference white (left) and zinc white varnished with mastic (right).

carried out, and those predicted with the model. The absolute and relative errors were also calculated, correlating the latter with the experiments' order of implementation (run) so as to discard any bias related to the way and order in which they were implemented. Equally, they were correlated with the value of each response, either in their observed values or in their predicted values, so as to also discard any possible bias [41].

MODDE Go® provides an indicator based on the relative weight of each factor or a combination of factors over all of the responses as a whole, called VIP (Variable Importance in Projection). In the oil painting cleaning model and as the most important factors, the following stand out in order of importance:

1. Water
2. Limonene
3. Pigments (lead white is the most important) and varnishes, and their synergies and antagonisms
4. Synergy-antagonism between water and limonene
5. Aging
6. Quadratic terms for water and limonene
7. Complex synergies and antagonisms between aging, pigments, varnish, water and limonene
8. Other components in the cleaner: Findet® 1214/N23, phenethyl alcohol and Glucopon® 600

The main factors that affect the cleaning process with the formulations proposed are water and limonene, as well as their synergies, antagonisms and quadratic terms. The concordance between the polarity of solvents and solutes is the fundamental matter in cleaning polychromies. Water acts as a modulator of polarity whereas limonene is a moderator of non-polarity, so their proportion in mixtures is decisive in the cleaning effect, as predicted by the model.

3.2 *In silico* cleaning simulations

After establishing the response surface model for cleaning varnishes on oil and having confirmed that the mathematical model is a good one, using the appropriate computer tools it is possible to carry out computer simulations, putting forward unlimited cleaning scenarios and analyzing them without having to carry them out physically. These types of techniques are often called "in silico", evoking the terms "in vivo" and "in vitro" common in the natural sciences and medicine.

The basis of these simulations has been created using the MODDE Go® 6.0 software, which allows triangular diagrams to be obtained that visually hold thousands of results in which all possible combinations of cleaner compositions have been simulated *in silico*, sweeping through all the ranges of concentrations of water, limonene, Findet® 1214/N23, Glucopon® 600 and phenethyl alcohol.

Figure 9 shows an example of a triangle diagram. In this example, the main solvents (limonene and water) and the main surfactant (Findet 1214/N23) are located at the vertices of each triangle. Within a triangle, there are colored areas corresponding to the different responses given by each cleaner depending on the type of varnish and pigment. Each level corresponds to the scale of values for the response in question: expert opinion with ultraviolet lighting and visible light; O/V cleaning with GC-ME; affectation from color as a distance, dELab, in the CIELAB space; affectation from lightness, ΔL; and affectation from gloss, ΔG.

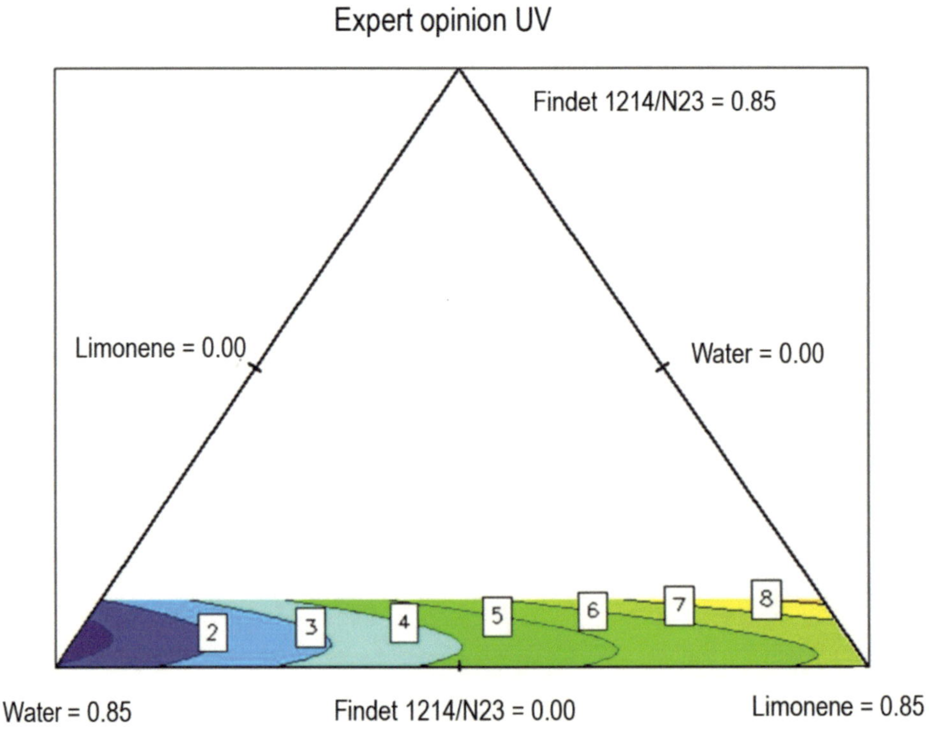

Figure 9.
Example of a triangular diagram of the results from cleaning with acrylic varnish according to expert opinion with ultraviolet light. Aging (Yes), Varnish (Acrylic). Pigment: Cadmium Red. Glucopon® 600: 0.1%. Phenethyl alcohol 0.25%.

The example in **Figure 9** comes from cleaning an oil painting made using cadmium red as a pigment and varnished with acrylic varnish that has undergone an aging process. The cleaners that have been simulated contain all the possible compositions of water, limonene and Findet® 1214/N23 (up to 10%) within the established ranges, in this case maintaining a fixed concentration of 10% of Glucopon® 600 (the maximum concentration established in designing experiments) and 2.5% of phenethyl alcohol (intermediate concentration in designing experiments). The response shown on the color scale is the expert's opinion using ultraviolet light, UV. To help with the analysis, the bottom section of the triangle is shown, where the cleaner's area of action is to be found, given the proportions of its three main components.

The simulations that have been carried out using this system are as follows:

1. Simulation in all ranges of the factors and for all responses. The results obtained are shown in triangular diagrams that enable a general "mapping" of the entire system to be obtained. This first approximation gives a general picture of the phenomenon of cleaning oil paintings, allowing us to visually find the main relationships between factors and responses.

2. Simulation for particular cases of cleaners that enable an evaluation of the complex relationships between the type of varnish, aging of the paint and type of pigment with the two main components of the cleaning compositions, water and limonene, using as a basis a cleaner with Findet® 1214/N23, Glucopon® 600 and phenethyl alcohol in fixed amounts (10%).

3. Tests to optimize the model in order to develop cleaning formulations with special characteristics that are optimum for performing their purpose.

4. Conclusions

1. A response surfaces model has been proposed for cleaning oil paintings with aqueous-based and limonene cleaning formulas using the MODDE Go® program, and its statistical validity has been demonstrated.

2. Thanks to the model, it has been possible to simulate a multitude of cleaning scenarios *in silico* and to determine the main factors that affect the cleaning, which is evaluated via the responses: O/V cleaning, expert opinion using visible and ultraviolet light, color affected, dELab, percentage of lightness affected, ΔL, and percentage of gloss affected, ΔG.

3. The main factors influencing the cleaning were the concentrations with water and limonene as the main solvents and which regulate the cleaners' level of hydrophilia and lipophilicity, followed by the type of varnish, aging and types of pigments. The cleaners' other components are less relevant. In decreasing order of relevance, they are Findet® 1214/N23, phenethyl alcohol and Glucopon® 600.

4. The cross-synergistic and antagonistic effects between the cleaners' components, the pigments, the varnish and the extent of aging have also been found to be very significant in cleaning.

5. Using *in silico* simulation, it is possible to formulate specific compositions for different scenarios and cleaning applications.

6. For future research, it is proposed to develop 3-dimensional vector models that include, firstly, the material and physico-chemical aspects of cleaning; secondly, factors related to the expert restorer; and thirdly, the visual dimension that can be evaluated with optical methods.

Acknowledgements

This research was funded by the following projects: *Desarrollo de Nuevas Sinergias Arte-Ciencia aplicadas a la Conservación y Restauración de los Palacios y Jardines de la Alhambra y el Generalife* (VIRARTE), MINECO, with reference HAR2016-79886-P; *Métodos sinérgicos Arte-Ciencia-Tecnología para la Conservación-Restauración de la Alhambra y otros Bienes Culturales* (VIRARTE II), MICINN, with referece PID2019-109713RB-I00; *La aplicación de las algas procedentes de la Alhambra y el Generalife en técnicas artísticas y de conservación-restauración* (FICOARTE), Universidad de Granada, with reference A-HUM-279-UGR18. "Aplicación avanzada de las algas procedentes de la Alhambra y el Generalife en técnicas artísticas y de conservación-restauración", (FICOARTE 2) (PAIDI 2020) with reference P18-FR-4477.

Author details

Julio Romero-Noguera[1*], Nuria Pérez-Villares[2], Fernando Bolívar-Galiano[2] and Rafael Bailón-Moreno[3]

1 Painting Department, Faculty of Fine Arts, University of Seville, Seville, Spain

2 Painting Department, Faculty of Fine Arts, University of Granada, Granada, Spain

3 Chemical Engineering Department, Faculty of Sciences, University of Granada, Granada, Spain

*Address all correspondence to: juliorn@us.es

IntechOpen

© 2022 The Author(s). Licensee IntechOpen. This chapter is distributed under the terms of the Creative Commons Attribution License (http://creativecommons.org/licenses/by/3.0), which permits unrestricted use, distribution, and reproduction in any medium, provided the original work is properly cited. (cc) BY

References

[1] Matteini M, Moles A. Science and Restoration. 2nd ed. San Sebastián: Nerea; 2002. p. 308

[2] Pinna D, Galeotti M, Mazzeo R. Scientific Examination for the Investigation of Paintings. A Handbook for Conservator-restorer. 1st ed. Firenze: Centro Di; 2009. p. 222

[3] Instituto de Patrimonio Histórico Español (IPHE). La Ciencia y el Arte: Ciencias Experimentales y Conservación del Patrimonio Histórico I-III. In: Madrid: Ministerio de Cultura. 1st ed. Subdirección General de Publicaciones, Información y Documentación; 2008. Available from: https://ipce.mecd.gob.es/difusion/publicaciones/libros-del-ipce.html

[4] Doménech Carbó MT. Análisis químico y examen científico de patrimonio cultural. 1st ed. Madrid: Síntesis; 2018. p. 334

[5] Artioli G. Scientific Methods and Cultural Heritage: An introduction to the application of Materials Science to Archaeometry and Conservation Science. 1st ed. Oxford: Oxford University Press; 2010. p. 554

[6] Feller RL. (Vol. I), Roy A (Vol. II), West Fitzhugh E (Vol. III), Berry B (Vol. IV). Artist's Pigments: A Handbook of Their History and Characteristics, Vol. I-IV. London: Archetype Publications; 2012

[7] Eastaugh N, Walsh V, Chaplin T, Siddall R. Pigment Compendium: A Dictionary and Optical Microscopy of Historical Pigments. 1st ed. London: Routledge; 2008. p. 968

[8] Gómez ML. La Restauración. In: Examen Científico Aplicado a la Conservación de Obras de Arte. 4th ed. Madrid: Cuadernos de Arte Cátedra; 2005. p. 488

[9] Moncrieff A, Weaver G. Conservation unit museums and galleries commission. The science for conservators series. In: Volumen 2. Cleaning. 2nd ed. London: Routedledge; 2005. p. 140

[10] Plenderleith HJ. The Conservation of Antiquities and Works of Art: Treatment, Repair and Restoration. 2nd ed. Oxford: Oxford University Press; 1971. p. 414

[11] Moncrieff A, Weaver G. Conservation unit museums and galleries commission. The science for conservators series. In: Volume 1: An Introduction to Materials. 2nd ed. London: Routedledge; 1992. p. 120

[12] Liants M-KL. Vernis et Adhésifs Anciens. 3rd ed. Bruxelles: Institut Royal du Patrimoine Artistique; 1992. p. 123

[13] Mayer R. The Artist's Handbook of Materials and Techniques. 5th ed. NYC: Viking Penguin Books; 1991. p. 784

[14] Cremonesi P, Bellucci R. L'uso dei Tensioattivi nella Conservazione en el restauro dei dipinti. Kermes. 1995;**24**:55-74

[15] Phenix A. 2002. The swelling of artists' paints in organic solvents. Part 1, A simple method for measuring the in-plane swelling of unsupported paint films. Journal of the American Institute for Conservation. 2002;**41**(1):43-60. DOI: 10.2307/3179896

[16] Guizzo S, Tortolini C, Pepi F, Leonelli F, Mazzei F. Di Turo F, Favero G. Application of microemulsions for the removal of synthetic resins from

paintings on canvas. Natural Product Research. 2019;**33**(7): 1015-1025. DOI:10.1080/14786419.2016.1244191

[17] Domingues JAL, Bonelli N, Giorgi R, Fratini E, Gorel F, Baglioni P. Innovative hydrogels based on semi-interpenetrating P(Hema)/PVP networks for the cleaning of water-sensitive cultural heritage artifacts. Langmuir. 2013;**29**(8):2746-2755. DOI: 10.1021/la3048664

[18] Bonelli N, Chelazzi D, Baglioni M, Giorgi R, Baglioni P. Confined aqueous media for the cleaning of cultural heritage: Innovative gels and amphiphile-based nanofluids. In: Dillmann P, Bellot-Gurlet L, Nenner I, editors. Nanoscience and Cultural Heritage. Paris: Atlantis Press; 2016. pp. 283-311. DOI: 10.2991/978-94-6239-198-7_10

[19] Erhardt D, Tumosa CS, Mecklenburg MF. Long-term chemical and physical processes in oil paint films. Studies in Conservation. 2005;**50**(2):143-150. DOI: 10.1179/sic.2005.50.2.143

[20] Phenix A, Burnstock A. The removal of surface dirt on paintings with chelating agents. The Conservator. 1992;**16**(1):28-38. DOI: 10.1080/01400096.1992.9635624

[21] Cremonesi P. L'uso dei SolventI Organici nella Pulitura di Opere Policrome. 1st ed. Villatora: Il Prato; 2004. p. 166

[22] Southall A. Detergents soaps surfactants. In: Dirt and pictures separated; UKIC and the Tate Gallery. January 1990; London. London: United Kingdom Institute for Conservation of Historic and Artistic Works. 1990;**1990**:29-34

[23] Kanjian H, Stulik D, Miller D. Research into solvent residues. In: In: Dorge V., editor. Solvents gels for the cleaning of works of art. The residue question. 1st ed. Los Ángeles: Getty Publications; 2004. pp. 66-83

[24] Romero-Noguera J, Bolívar-Galiano FC, Ramos-López JM, Fernández-Vivas MA, Martín-Sánchez I. Study of biodeterioration of diterpenic varnishes used in art painting: Colophony and Venetian turpentine. International Biodeterioration and Biodegradation. 2008;**62**(4):427-433. DOI: 10.1016/j.ibiod.2008.03.014

[25] Romero-Noguera J, Martín-Sánchez I, López-Miras MM, Ramos-López JM, Bolívar-Galiano FC. Biodeterioration patterns found in dammar resin used as art material. Electronic Journal of Biotechnology. 2009;**13**(3). DOI: 10.2225/vol13-issue3-fulltext-7

[26] Romero-Noguera J, López-Miras MM, Martín-Sánchez I, Ramos-López JM, Bolívar-Galiano FC. An approach to the study of the fungal deterioration of a classical art material: Mastic varnish. Electronic Journal of Biotechnology. 2010;**13**(6). DOI: 10.2225/vol13-issue6-fulltext-11

[27] Romero-Noguera J, Doménech-Carbó MT, Osete-Cortina L, Martín-Sánchez I, Bolívar-Galiano FC. Analytical characterisation of the biodeterioration of diterpenoid labdanic varnishes used in pictorial techniques: Sandarac and Manila copal. International Biodeterioration and Biodegradation. 2014;**90**:99-105. DOI: 10.1016/j.ibiod.2014.01.017

[28] Doerner M. The Materials of the Artist and their use in Painting: With Notes on the Techniques of the Old Masters. Revised ed. Boston: Houghton Mifflin; 1949. p. 468

[29] Van der Weerd J, Van Loon A, Boon JJ. Estudios FTIR de los efectos de

los pigmentos sobre el envejecimiento del aceite. Studies in Conservation. 2005;**50**(1):3-22. DOI: 10.1179/sic.2005.50.1.3

[30] Hermans JJ. Metal Soaps in Oil Paint: Structure, Mechanisms and Dynamics [thesis]. Amsterdam: Van 't Hoff Institute for Molecular Sciences (HIMS). 2017;**2017**:162

[31] Matteini M, Mazzeo R, Moles A. Chemistry for Restoration. Painting and restoration Materials. 1st ed. Firenze: Nardini Editore; 2016. p. 372

[32] Mills JS, White R. The Organic Chemistry of Museum Objects. 2nd ed. London: Routledge; 1999. p. 222

[33] Horie V. Materials for Conservation: Organic Consolidants, Adhesives and Coatings. Revised ed. London: Routledge; 2010. p. 504

[34] Carlson J, Petersen C. Aging characteristics of surfactants. In: In: Dorge V., editor. Solvents Gels for the Cleaning of Works of Art. The Residue Question. 1st ed. Los Ángeles: Getty Publications; 2004. pp. 84-111

[35] Cremonesi P. L'ambiente Acquoso per la Pulitura di Opere Policrome. 1st ed. Villatora: Il Prato; 2011. p. 112

[36] Materiales CA. Técnicas y Procedimientos. De la A a la Z. 2nd ed. Barcelona: Serbal; 2006. p. 256

[37] Wolbers R. Cleaning Painted Surfaces: Aqueous Methods. 1st ed. London: Archetype Books; 2000. p. 198

[38] Torraca G. Solubility and Solvents for Conservation Problems. 4th Edition, reprinted ed. Roma: ICCROM; 2005. p. 76

[39] Feller R, Stolow N, Jones EH. On picture Varnishes and their Solvents. Revised and Enlarged ed. Washington: National Gallery of Art; 1985. p. 260

[40] Hansen CM. Hansen Solubility Parameters: A user's handbook. 2nd ed. Boca Raton: CRC Press; 2007. p. 546

[41] Bailón R, Romero-Noguera J, Bolívar FC, Yebra AM, Pérez N. Response surfaces model for restoring and cleaning oil painted artworks. Journal of Cultural Heritage. 2020;**45**:10-24. DOI: 10.1016/j.culher.2020.03.011

[42] Pérez-Villares N, Bailón-Moreno R. New simple procedure to produce white lead for special use in the plastic arts and in restoration. Journal of Cultural Heritage. 2017;**27**:164-169. DOI: 10.1016/j.culher.2016.11.001

[43] Wyszecki G. Colour Science. Concepts and Methods. Quantitative Data and Formulae. 2nd ed. Hoboken: John Wiley and Sons; 2021. p. 950

[44] Völz HG. Industrial Colour Testing. Fundamentals and Techniques. 2nd ed. Hoboken: John Wiley and Sons; 2003. p. 388